T0282622

CAMBRIDGE LIBRARY COLLECTION

Books of enduring scholarly value

Botany and Horticulture

Until the nineteenth century, the investigation of natural phenomena, plants and animals was considered either the preserve of elite scholars or a pastime for the leisured upper classes. As increasing academic rigour and systematisation was brought to the study of 'natural history', its subdisciplines were adopted into university curricula, and learned societies (such as the Royal Horticultural Society, founded in 1804) were established to support research in these areas. A related development was strong enthusiasm for exotic garden plants, which resulted in plant collecting expeditions to every corner of the globe, sometimes with tragic consequences. This series includes accounts of some of those expeditions, detailed reference works on the flora of different regions, and practical advice for amateur and professional gardeners.

Flowers of the Field

A keen collector and sketcher of plant specimens from an early age, the author, educator and clergyman Charles Alexander Johns (1811–74) gained recognition for his popular books on British plants, trees, birds and countryside walks. *The Forest Trees of Britain* (1847–9), one of several works originally published by the Society for Promoting Christian Knowledge, is also reissued in this series. First published by the Society in 1851, Johns' best-known work is this two-volume botanical guide to common British flowering plants. Following the Linnaean system of classification, Johns describes the various plant families, providing the common and Latin names for each species. The work is especially noteworthy for its delicate and meticulous line drawings, based on watercolours by the botanical artist Emily Stackhouse and the author's sisters Julia and Emily. Volume 2 completes the botanical descriptions and includes indexes of English and Latin plant names.

Cambridge University Press has long been a pioneer in the reissuing of out-of-print titles from its own backlist, producing digital reprints of books that are still sought after by scholars and students but could not be reprinted economically using traditional technology. The Cambridge Library Collection extends this activity to a wider range of books which are still of importance to researchers and professionals, either for the source material they contain, or as landmarks in the history of their academic discipline.

Drawing from the world-renowned collections in the Cambridge University Library and other partner libraries, and guided by the advice of experts in each subject area, Cambridge University Press is using state-of-the-art scanning machines in its own Printing House to capture the content of each book selected for inclusion. The files are processed to give a consistently clear, crisp image, and the books finished to the high quality standard for which the Press is recognised around the world. The latest print-on-demand technology ensures that the books will remain available indefinitely, and that orders for single or multiple copies can quickly be supplied.

The Cambridge Library Collection brings back to life books of enduring scholarly value (including out-of-copyright works originally issued by other publishers) across a wide range of disciplines in the humanities and social sciences and in science and technology.

Flowers of the Field

VOLUME 2

CHARLES ALEXANDER JOHNS

CAMBRIDGE
UNIVERSITY PRESS

CAMBRIDGE
UNIVERSITY PRESS

University Printing House, Cambridge, CB2 8BS, United Kingdom

Published in the United States of America by Cambridge University Press, New York

Cambridge University Press is part of the University of Cambridge.
It furthers the University's mission by disseminating knowledge in the pursuit of
education, learning and research at the highest international levels of excellence.

www.cambridge.org
Information on this title: www.cambridge.org/9781108068659

© in this compilation Cambridge University Press 2014

This edition first published 1853
This digitally printed version 2014

ISBN 978-1-108-06865-9 Paperback

FLOWERS OF THE FIELD.

BY THE

REV C. A. JOHNS, B.A. F.L.S.

AUTHOR OF BOTANICAL RAMBLES, THE FRUIT TREES OF BRITAIN,
ETC. ETC.

IN TWO VOLUMES.
VOL. II.

PUBLISHED UNDER THE DIRECTION OF
THE COMMITTEE OF GENERAL LITERATURE AND EDUCATION,
APPOINTED BY THE SOCIETY FOR PROMOTING
CHRISTIAN KNOWLEDGE.

LONDON:
PRINTED FOR THE
SOCIETY FOR PROMOTING CHRISTIAN KNOWLEDGE;
SOLD AT THE DEPOSITORY,
GREAT QUEEN STREET, LINCOLN'S INN FIELDS,
4, ROYAL EXCHANGE, 16, HANOVER STREET, HANOVER SQUARE;
AND BY ALL BOOKSELLERS.

LONDON:
R. CLAY, PRINTER, BREAD STREET HILL.

NATURAL ARRANGEMENT OF PLANTS.

ORD. XLVII.—CAMPANULACEÆ.—THE BELL-FLOWER TRIBE.

Calyx growing from the ovary, 5-lobed, remaining till the fruit ripens; *corolla* of one petal, rising from the mouth of the calyx, 5-lobed, regular, withering on the fruit; *stamens* equalling in number the lobes of the corolla, and alternate with them; *anthers* not united (except in *Jasioné*); *ovary* inferior, of 2, or more, many-seeded cells; *style* 1, covered with hairs; *stigma* simple, or with as many lobes as the ovary has cells; *fruit* dry, crowned by the withered calyx and corolla, splitting, or opening by valves, at the side or top; *seeds* numerous, fixed to a central column.—Herbaceous or slightly shrubby plants, with a milky bitter juice, mostly alternate leaves without stipules, and showy blue or white flowers, inhabiting principally the temperate regions of the northern hemisphere. Many species are highly ornamental, but very few are valuable either as food or medicine. The roots of *Campánula Rapúnculus*, under the name of Rampion or Ramps,

were formerly cultivated in this country for the table,
but are now scarcely known.

1. CAMPÁNULA (Bell-flower).—*Corolla* bell-shaped,
with 5 broad and shallow lobes; *filaments* broad at the
base; *stigma* 2—5 cleft; *capsule* 2—5 celled, opening
by pores at the side. (Name from the Latin *campana*,
a bell.)

2. SPECULARIA.—*Corolla* wheel-shaped, with 5 lobes;
filaments broad at the base; *capsule* oblong, triangular,
opening by pores near the top. (Name from *speculum*,
a prism, from the prismatic or triangular shape of the
capsule.)

3. PHYTEUMA (Rampion).—*Corolla* wheel-shaped,
with 5 deep lobes; *filaments* broad at the base; *stigma*
2—3 cleft; *capsule* 2—3 celled, bursting at the side.
(Name from the Greek *phyton*, a plant.)

4. JASÍONÉ (Sheep's-Scabious).—*Corolla* wheel-shaped,
with 5 long narrow segments; *anthers* united at the
base; *stigma* 2-cleft; *capsule* 2-celled, opening at the
top; *flowers* in heads. (Name of uncertain origin.)

1. CAMPANULA (*Bell-flower*).

1. *C. rotundifolia* (Hair-bell).—Smooth; *root-leaves*
roundish kidney-shaped, notched, stalked, very soon
withering; *stem-leaves* very narrow, tapering.—Heaths
and dry meadows, abundant. The name Hair-bell is
frequently, though not correctly, given to the Wild
Hyacinth or Blue-bell (*Scilla nutans*, or *Hyacinthus
non-scriptus*), a plant with a thick juicy flower-stalk;
but when applied to this *Campánula* is most appro-
priate, its stalks being exceedingly slender and wiry.
The specific name, *rotundifolia* (round-leaved), is far
from being descriptive of the leaves which accompany
the flower, as they are long and narrow, but is peculiarly
applicable to the root leaves, as they appear in winter
or early spring, at which season Linnæus is said to have

first observed them on the steps of the university at
Upsal.—Fl. July—September.   Perennial.

CAMPANULA ROTUNDIFOLIA (*Hair-bell*).

2. *C. Trachelium* (Nettle-leaved Bell-flower).—*Lower
leaves* stalked, heart-shaped ; *upper* nearly sessile, tapering to a sharp point, all strongly serrated and bristly;
flowers in axillary clusters of 2—3.—Woods and hedges,
not unfrequent.   A remarkably rough plant, 2—3 feet
high, with leaves very like those of the nettle, and large,
deep blue, bell-shaped flowers, the stalks of which are
recurved when in fruit.—Fl. July, August.   Perennial.

3. *C. glomerata* (Clustered Bell-flower).—*Stem* simple, roughish ; *leaves* oblong, tapering, crenate, rough, the lower stalked and heart-shaped at the base, the upper sessile, embracing the stem ; *flowers* sessile, in heads.— Dry pastures, not unfrequent. A stiff, erect plant, 3 — 18 inches high, with terminal and (in large specimens) axillary heads of deep blue, funnel-shaped, erect flowers, which have a few clasping, taper-pointed bracts at the base.—Fl. July, August. Perennial.

4. *C. hederacea* (Ivy-leaved Bell-flower).—*Stem* straggling, thread-like ; *leaves* stalked, roundish heart-shaped, angular, and toothed ; *flowers* solitary, on long stalks.—Wet heaths and by the sides of streams in the south and west ; very abundant in Cornwall. An exquisite little plant, generally growing with *Anagallis tenella*, and (in Cornwall) with *Sibthorpia Europœa*, plants certainly of a different habit, but scarcely less elegant than itself. The leaves are of a remarkably fine texture, and delicate green hue ; the flowers of a pale blue, sometimes slightly drooping, and supported on long stalks scarcely thicker than a hair. Its usual height is 4—6 inches, but when it grows among grass or rushes, it climbs by their help to a height of 12 inches or more.—Fl. July—September. Perennial.

* Less common species of *Campánula* are *C. pátula* (Spreading Bell-flower), distinguished by its rough *stem*, and loose *panicles* of wide cup-shaped *flowers: C. Rapúnculus* (Rampion Bell-flower), a tall species, 2—3 feet high, with clustered *panicles* of rather small, pale blue *flowers*, the *calyx* of which is divided into 5 awl-shaped segments : *C. latifolia* (Giant Bell-flower), common in woody glens in Scotland, but not frequent in England ; a stout species, 3—4 feet high, with very large stalked *flowers*, which are deep blue, and hairy within : and *C. Rapunculoídes* (Creeping Bell-flower), a very rare species, distinguished by its pale blue, drooping, axillary *flowers*, which grow all on one side of the stem.—Fl. July, August. Perennial.

2. Specularia.

1. *S. hýbrida.*—The only British species.—A small plant, 4—12 inches high, with a rough wiry stem, oblong, rough, wavy *leaves*, and a few small terminal purple *flowers*, the *calyx* of which is much longer than the *corolla.*—Corn-fields, not common. By some botanists this is called *Campánula hýbrida* (Corn Bell-flower).—Fl. June—September. Annual.

3. Phyteuma (*Rampion*).

1. *P. orbiculáré* (Round-headed Rampion).—*Flowers* in a round terminal head ; *lower leaves* notched, heart-shaped, stalked ; *upper* narrow, sessile.—Chalky downs, rare. A singular plant, consisting of a solitary, erect leafy stalk, 12—18 inches high, surmounted by a round head of blue flowers. The head when in fruit becomes oval.—Fl. July. Perennial.

* *P. spicatum* (Spiked Rampion) is found only in Sussex. It is much taller than the last, and bears its *flowers*, which are cream-coloured, in a terminal, oblong head.

4. Jasíoné (*Sheep's Scabious*).

1. *J. montána* (Sheep's Scabious, Sheep's-bit).—The only British species. — Dry heathy places, common. Growing about a foot high, and having a strong resemblance to a Scabious, from which it may be at once distinguished by its united *anthers ;* or to a Compound Flower, from which it differs in having a 2-celled capsule. The *leaves* are oblong, blunt, and hairy ; the *flowers*, which are blue, grow in terminal *heads*, with a leafy *involucre* at the base. The whole plant, when

bruised, has a strong and disagreeable smell.—Fl. July,
August. Biennial.

JASIONE MONTANA (*Sheep's Scabious, Sheep's-bit*).

Ord. XLVIII.—LOBELIACEÆ.—The Lobelia Tribe.

Calyx growing from the ovary, 5-lobed, or entire;
corolla of one petal, inserted in the calyx, 5-lobed, irregular; *stamens* equalling in number the lobes of the
corolla, and alternate with them; *anthers* united; *ovary*

inferior, of 1—3 many-seeded cells, opening at the top.
—The plants comprised in this Order resemble in many
respects the Bell-flower Tribe, from which they are

LOBELIA DORTMANNA (*Water Lobelia*).

mainly distinguished by their irregular corolla and
united anthers. Like them, they contain a milky juice,
which, however, is more acrid, and they inhabit generally
warmer regions. *Lobelia inflata* (Indian Tobacco) pos-

sesses powerful medicinal properties, and when given in
over-doses is poisonous. *L. cardinalis* (Scarlet Cardinal), one of our most brilliantly-coloured garden flowers,
is also very acrid ; and the rare British species, *L. úrens*
(Acrid Lobelia), derives its name from the blistering
properties of its juice. Some species contain a considerable quantity of caoutchouc.

1. LOBELIA.—*Corolla* 2-lipped, the upper part split
to the base of the tube. (Name from Matthias Lobel, a
Flemish botanist.)

1. LOBELIA.

1. *L. Dortmanna* (Water Lobelia).—*Leaves* almost
cylindrical, of 2 parallel tubes.—Lakes in the north,
frequent. An aquatic plant, often forming a matted
bed at the bottom of the water, and sending above the
surface slender, almost leafless, stems, having a long
cluster of distant, light blue, drooping flowers.—Fl. July,
August. Perennial.

* *L. úrens* (Acrid Lobelia) is a very rare species,
found only near Axminster, Devon ; it has a roughish
leafy *stem*, which contains a milky, acrid juice, and
leafy clusters of purple *flowers*.

ORD. XLIX.—VACCINIEÆ.—THE CRANBERRY TRIBE.

Calyx growing from the ovary, of 4—6 lobes, which
are sometimes so shallow as to be scarcely perceptible ;
corolla of one petal, with as many lobes as the calyx ;
stamens not united, twice as many as the lobes of the
corolla, inserted into the *disk* of the ovary ; *anthers* opening by 2 pores, and often furnished with 2 bristles ;
ovary with a flat disk, 4—10 celled ; *cells* 1 or many,
seeded ; *style* and *stigma* simple ; *fruit* a *berry* crowned
by the remains of the calyx, juicy, containing many
small *seeds*.—Small shrubby plants, with undivided,
alternate leaves, inhabiting temperate regions, especially

mountainous and marshy districts. By some botanists they are placed in the same order with the Heaths, from which they differ chiefly in having the ovary beneath the calyx. The leaves and bark are astringent, the berries slightly acid and agreeable to the taste. Under the name of Cranberries, they are imported largely from North America, and are used for making tarts. Many species are cultivated in gardens; more, however, for their pretty flowers than for the sake of their fruit.

1. VACCINIUM (Whortleberry, Cranberry, &c.)—*Calyx* 4—5 lobed, sometimes with the lobes so shallow as to be scarcely perceptible ; *corolla* bell-shaped, or wheel-shaped, 4—5 cleft ; *stamens* 8—10 ; *berry* globose, 4—5 celled, many seeded. (Name of doubtful etymology.)

VACCINIUM (*Whortleberry, Cranberry, &c.*)

* *Leaves not evergreen ; anthers with two bristles at the back.*

1. *V. myrtillus* (Whortleberry, Bilberry, Whinberry). —*Stem* acutely angular ; *leaves* egg-shaped, serrated ; *flowers* solitary, drooping. — Heathy and mountainous places ; abundant. A small branched shrub, 6—18 inches high, with nearly globular, flesh-coloured, waxy *flowers*, and black *berries*, which are covered with grey bloom. They are agreeable to the taste, and are often made into tarts; but when thus used are rather mawkish unless mixed with some more acid fruit. In the west of England they are popularly known by the name of *whorts*. Fl. May. Shrub.

2. *V. uliginosum* (Bog Whortleberry, or Great Bilberry).—*Stem* not angular ; *leaves* inversely egg-shaped, entire, glaucous and veined beneath.—Mountainous bogs in Scotland and the north of England. Distinguished from the last by its more woody, rounded, stem, and by its strongly veined, glaucous leaves, which are broader

towards the extremity. The flowers are smaller and grow nearer together. Fl. May. Shrub.

VACCINIUM MYRTILLUS (*Whortleberry, Bilberry, Whinberry*).

** *Leaves evergreen ; anthers without bristles.*

3. *V. Vitis Idæa* (Red Whortleberry, Cowberry).— *Leaves* inversely egg-shaped, dotted beneath, the margins

rolled back ; *flowers* in terminal drooping clusters.—
Mountainous heaths in the north. A low straggling
shrub, with leaves resembling those of the *Box.* The
flowers are pink, with 4 deep lobes ; the berries red.
Fl. May, June. Shrub.

4. *V. Oxycoccos* (Marsh Whortleberry, Cranberry).—
Stem very slender, prostrate, rooting ; *leaves* egg-shaped,
glaucous beneath, the margins rolled back ; *corolla* wheel-
shaped, with 4 deep, reflexed segments.—Peat-bogs,
principally in the north. A very low plant with straggling, wiry stems, and solitary, terminal, bright red
flowers, the segments of which are bent back in a very
singular manner. " The fruit is highly agreeable,
making the best of tarts ; at Langtown, on the borders
of Cumberland, it forms no inconsiderable article of
trade." (*Sir W. J. Hooker.*) Fl. June. Shrub.

Ord. L.—ERÍCEÆ.—The Heath Tribe.

Calyx 4—5 cleft, nearly equal, inferior, remaining
till the fruit ripens ; *corolla* of one petal 4—5 cleft, often
withering and remaining attached to the plant ; *stamens*
equal in number to the segments of the corolla, or twice
as many, inserted with the corolla or only slightly
attached to its base ; *anthers* hard and dry, the cells
separating at one extremity, where they are furnished
with bristles or some other appendage, opening by pores :
ovary not adhering to the calyx, surrounded at the base
by a *disk* or by *scales,* many-celled, many-seeded ; *style*
1, straight ; *stigma* 1 ; *fruit* a berry or dry capsule,
many seeded.—Shrubs or small bushy trees with evergreen, often rigid, opposite or whorled leaves. This
well-known and highly-prized order contains a large
number of beautiful plants, many of which are remarkable for their social qualities ; extensive tracts of country being often found entirely covered with a few species,
so as to give name to the kinds of places on which they

grow.   They are most abundant in the neighbourhood
of the Cape of Good Hope ; hence they are often called
by gardeners, "Cape plants."   They are common also in
Europe, in North and South America, both within and

ERICA TETRALIX, E. CILIARIS, E. VAGANS, *and* E. CINEREA.

without the tropics, and in the mountainous parts of
Asia.   The extensive genus *Eríca* (Heath) contains no
plant possessing useful properties.   *Callúna vulgáris*
(Ling, or Heather,) is astringent, and is sometimes used
for dyeing ; its rough branches are a common material for
brooms ; its flowers are the favourite resort of bees, and
its seeds are said to be the principal food of moor-fowl.

Of the plants belonging to the order, which produce juicy berries, the fruit is in some instances edible. *Árbutus Únedo* bears an abundance of handsome berries, which when thoroughly ripe are not unpalatable; and which, from the resemblance they outwardly bear to strawberries, give the plant its English name. Some species, especially *Kalmia* and *Rhododendron,* possess dangerous narcotic properties, which extend to the flesh of animals that have fed on them. It is stated that the honey which poisoned the Grecian troops during the famous Retreat of the 10,000, had been collected by bees from the flowers of some plant of this order; and that the honey still found on the shores of the Euxine, or Black Sea, possesses the same properties. The berries of some species are, nevertheless, used in medicine with good effect.

1. Eríca (Heath).—*Calyx* deeply 4-cleft; *corolla* bell-shaped, or egg-shaped, 4-cleft; *stamens* 8; *capsule* 4-celled. (Name from the Greek *eríco,* to break, from some fancied medicinal properties.)

2. Callúna (Ling, Heather).—*Calyx* of 4 coloured sepals, which are longer than the corolla, having at the base outside 4 green *bracts; corolla* bell-shaped; *stamens* 8; *capsule* 4-celled. (Name, from the Greek *callúno,* to cleanse, from the frequent use to which its twigs are applied, of being made into brooms.)

3. Menziesia.—*Calyx* deeply 4—5 cleft; *corolla* inflated; *stamens* 8—10; *capsule* 4—5 celled. (Named in honour of *Archibald Menzies,* an eminent Scotch botanist.)

4. Azálea.—*Calyx* deeply 5-cleft; *corolla* bell-shaped, 5 cleft; *stamens* 5; *anthers* bursting lengthways; *capsule* 2—3 celled, and valved. (Name from the Greek *azáleos,* parched, from the nature of the places in which it grows.)

5. Andrómeda.—*Calyx* deeply 5-cleft; *corolla* egg-shaped, with a 5-cleft reflexed border; *stamens* 10; *anthers* with 2 bristles at the back; *capsule* dry, 5-celled

and 5-valved. ("Named in allusion to the fable of
Andrómeda, who was chained to a rock, and exposed
to the attack of a sea-monster : so does this beautiful
tribe of plants grow in dreary and northern wastes,
feigned to be the abode of preternatural monsters."—
Sir W. J. Hooker.)

6. ÁRBUTUS (Strawberry-tree, Bear-berry).—*Calyx*
deeply 5-cleft ; *corolla* egg-shaped, with a 5-cleft reflexed
border ; *stamens* 10 ; *fruit* a 5-celled, many-seeded
berry. (Name, the Latin name of the plant.)

1. ERÍCA (*Heath*).

1. *E. Tétralix* (Cross-leaved Heath).—*Leaves* 4 in a
whorl, narrow, fringed ; *flowers* in 1-sided terminal
heads.—Peaty moors ; abundant ; well distinguished
from all other English species by its leaves being placed
cross-wise, and by its terminal heads of drooping rose-
coloured flowers, which are all turned to the same side,
and are of a larger size than the other common species,
E. cinérea. The part of the flower nearest the stem
is of a lighter colour than that which is exposed, where
it deepens to a delicate blush ; the whole flower appearing as if it had been modelled in wax. It is sometimes
found of a pure white. Fl. July, August, with occa
sional blooms throughout the autumn. Shrub.

2. *E. cinérea* (Fine-leaved Heath).—*Leaves* in threes,
very narrow, smooth ; *flowers* egg-shaped, in irregular,
whorled, leafy clusters.—Heaths, abundant. This and
the preceding are the only *Heaths* which can be called
common. It is a bushy plant, with tough, wiry stems,
exceedingly narrow leaves, and numerous oblong, purple flowers, which form broken, leafy clusters, not confined to one side of the stem. The flowers are sometimes
white.—Fl. July, August. Shrub.

3. *E. vagans* (Cornish Heath).—*Leaves* 3—5 in a
whorl, crowded, very narrow, smooth ; *flowers* bell-
shaped, shorter than the stamens, forming a leafy, regular, tapering cluster.—Heaths on the southern promon-

tory of Cornwall, very abundant ; found also in one or two other places in Cornwall, and on the coast of Waterford. Stems much branched, and, in the upper

ERICA TETRALIX (*Cross-leaved Heath*).

parts, very leafy, 2—4 feet high : flowers light purple, rose-coloured, or pure white. In the purple variety the anthers are dark purple ; in the white, bright red ; and in all cases they form a ring outside the corolla until they have shed their pollen, when they droop to the sides. On the Goonhilley Downs, in Cornwall, all these varieties of this Heath grow together in the greatest

profusion, covering many hundreds of acres, and almost
excluding the two species so common elsewhere.—Fl.
July—September.   Shrub.

ERICA VAGANS (*Cornish Heath*).

   * Less common species of Heath are *E. Mackáii*
(Mackay's Heath), approaching closely to *E. Tétralix*,
but growing in a more bushy manner, with broader
leaves, and more numerous heads of smaller *flowers;*
Cunnemara, Ireland. *E. Mediterranea* (Mediterranean
Heath), well known as a garden plant, but found, also,
in Cunnemara, well marked by its coloured calyx ; and
E. ciliaris (Ciliated Heath), by far the most beautiful

of all the British species ; the *leaves* are 4 in a whorl, and the *flowers*, which are bright purple, and half an inch long, grow in terminal, interrupted, spike-like clusters. It is found only near Cape Castle, Dorset, and in Cornwall, where, though of local occurrence, it is occasionally as abundant as *E. cinérea* is elsewhere.

CALLUNA VULGARIS (*Ling, or Heather*).

2. CALLÚNA (*Ling, Heather*).

1. *C. vulgáris* (Ling, or Heather).—The only species. Heaths and moors ; most abundant. This was placed

by Linnæus in the genus *Eríca* (Heath) ; later botanists
have, however, made of it a distinct genus, and not
without reason. The leaves are very small, more or
less downy (sometimes even hoary), and arranged in
4 rows, on opposite sides of the stem. The *corolla* is
very small, and bell-shaped, and is concealed by the
rose-coloured leaves of the *calyx*, outside which are
4 small green *bracts*, which appear to form a second
calyx. The flowers remain attached to the plant long
after the seed is ripe ; indeed, it is not at all unusual
to find plants in full bloom with the withered flowers of
the preceding year still adhering to the lower part of
the stem. A beautiful variety has been found in Cornwall, with double flowers.—Fl. July, August. Shrub.

3. MENZIESIA.

1. *M. cœrúlea* (Scotch Menziesia).—*Leaves* numerous, linear, minutely toothed ; *flower-stalks* covered
with glandular hairs ; *flowers* in terminal tufts ; *corolla*
5-cleft ; *stamens* 10.—Very rare ; found on the " Sow
of Athol," in Perthshire, but " nearly, if not quite
extirpated by an Edinburgh nurseryman."—*Babington.*
A small shrubby plant, naked below, very leafy and
hairy above, with large, pale purplish-blue flowers.—
Fl. June, July. Shrub.
2. *M. polifolia* (Irish Menziesia, or St. Dabeoc's
Heath).—*Leaves* egg-shaped, with the margins rolled
back, white, and downy beneath ; *corolla* 4-cleft ; *stamens* 8.—Mountainous heaths in Ireland ; rare. A small
shrub, with large, purple, sometimes white flowers, which
grow in terminal, leafy, 1-sided clusters.—Fl. June,
July. Shrub.

4. AZÁLEA.

1. *A. procumbens* (Trailing Azalea).—A low trailing
shrub, of a very different habit from the garden plants
cultivated under the name of *Azaleas.* The *stems* are
prostrate and tangled ; the *leaves* small, smooth, and

rigid, with the margins remarkably rolled back ; the *flowers* are flesh-coloured, and grow in short terminal clusters, or tufts. Highland mountains.—Fl. May, June. Shrub.

5. ANDRÓMEDA.

1. *A. polifolia* (Marsh Andrómeda).—The only British species, growing in peat bogs in the north. A small, leafy, evergreen shrub, with slender *stems,* narrow, pointed *leaves,* and terminal tufts of flesh-coloured, drooping *flowers.*—Fl. June—August. Shrub.

ARBUTUS UNEDO (*Strawberry-tree*).

6. ÁRBUTUS (*Strawberry-tree, Bear-berry*).

1. *A. Únedo* (Strawberry-tree). — *Leaves* elliptical, tapering, serrated, smooth ; *flowers* in drooping panicles ; *fruit* rough.—Abundant about the lakes of Killarney, in a wild state, and very common in English gardens.

A beautiful evergreen tree, with a rough, reddish bark, large, deep-green leaves, and numerous terminal clusters of greenish-white *flowers*. The *berries*, which ripen in the following autumn, are nearly globular, scarlet, and rough, with minute, hard grains. They are eatable, but so much less attractive to the taste than to the eye, as to have originated the name, " Unedo," " One-I-eat ;" as if no one would choose to try a second. The flowers are in full perfection at the time when the fruit, formed in the preceding year, is ripening ; and then, of course, the tree presents its most beautiful appearance.—Fl. September, October. Tree.

2. *A. Uva-Ursi* (Red Bear-berry).—*Stems* prostrate ; *leaves* inversely egg-shaped, entire, evergreen ; *flowers* in terminal clusters.—Mountainous heaths in the north; abundant. A small shrub, distinguished by its long trailing *stems*, blunt *leaves*, which turn red in autumn, rose-coloured *flowers*, and small red *berries*. The leaves are used in medicine as an astringent.—Fl. May. Shrub.

* *A. alpína* (Black Bear-berry), resembles the last in its mode of growth, but the *leaves* are wrinkled and serrated, and not evergreen ; the *flowers* are white, with a purplish tinge ; the *berries* black. It is most common on mountains in the north of Scotland.

Ord. LI.—MONOTROPEÆ.—The Bird's-nest Tribe.

Sepals 4—5, not falling off ; *corolla* regular, deeply divided into as many lobes or petals as there are sepals ; *stamens* twice as many as the lobes of the corolla ; *anthers* opening by pores ; *ovary* 4—5 celled, sometimes imperfectly so ; *style* 1, often bent ; *stigma* generally lobed ; *fruit* a dry capsule ; *seeds* covered with a loose skin.—A small and unimportant Order, containing but two British genera, *Pýrola*, a family of plants with somewhat shrubby, unbranched *stems;* simple, smooth, veiny, evergreen *leaves*, and large, often fragrant, *flowers*, which grow either singly or in a stalked terminal *cluster :* and *Monótropa*, a leafless parasitic plant, with the habit of an *Orobanché* (Broom-rape), growing on the roots of Firs, and other trees.

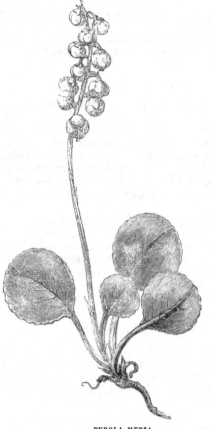

PYROLA MEDIA.

1. PÝROLA (Winter-green).—*Sepals* 5 ; *corolla* of 5 deep lobes, or petals ; *stamens* 10 ; *anthers* 2-celled ;

stigma 5-lobed. (Name signifying *a little Pear*, from the fancied resemblance between its leaves and those of that tree.)

2. Monótropa (Bird's-nest).—*Sepals* 4—5 ; *petals* 4—5, swollen at the base ; *stamens* 8—10 ; *anthers* 1-celled ; *stigma* flat, not lobed. (Name from the Greek *monos*, one, and *trepo*, to turn, the flowers being turned all one way.)

1. Pýrola (*Winter-green*).

1. *P. uniflóra* (Single-flowered Winter-green). — *Leaves* nearly round ; *flower* solitary, drooping.—Mountainous woods in Scotland ; rare. A remarkably pretty plant, and not less singular than beautiful. The lower part of the stem bears several roundish, egg-shaped, smooth, and veiny leaves, and runs up into a single flower-stalk, which is 3—6 inches high, and bears one large, elegant, white, highly fragrant flower.—Fl. July. Perennial.

* The other species of Winter-green, which bear each a terminal cluster of drooping flowers, though very different from the preceding species, are difficult to be distinguished from one another, especially when dried, in which state only they are to be seen in South Britain. *P. rotundifolia*, *P. minor*, and *P. secunda*, all approach in habit to the one figured, *P. media*, which is the most frequent.

2. Monótropa (*Bird's-nest*).

1. *M. Hypópitys* (Pine Bird's-nest, Fir-rape).—The only British species, occurring sparingly in dry woods of Fir and Beech, on the roots of which trees it is said by some to be parasitical. The whole plant consists of a single juicy stalk, without leaves, but clothed throughout with scaly *bracts*, and terminating in a drooping cluster of brownish yellow *flowers*, which eventually turn almost black. This must not be confounded with the plants of the genus *Orobanché*, which all have a ringent corolla

of 1 petal, and four stamens, two of which are shorter than the others. The flowers of *Monótropa* have 8 stamens, with the exception of the terminal one, which has 10.—Fl. June, July. Perennial.

MONOTROPA HYPOPITYS (*Pine Bird's-nest*, *Fir-rape*).

Ord. LII.—ILICINEÆ.—The Holly Tribe.

Sepals 4—6, imbricated when in bud ; *corolla* 4—6
lobed, imbricated when in bud ; *stamens* inserted into
the corolla, equalling its lobes in number, and alternate
with them ; *filaments* erect ; *anthers* 2-celled, opening
lengthwise ; *ovary* fleshy, abrupt, 2—6 celled ; *stigma*
nearly sessile, lobed ; *fruit* a fleshy berry, not bursting,
containing 2—6 bony seeds.—Evergreen trees or shrubs,
with tough leaves, and small axillary, white, or greenish
flowers, occurring in various parts of the world ; the
only European species being the common Holly. Nearly
all the plants of this tribe possess astringent and tonic
properties. The leaves of the Holly, for instance, are
said to be equal to Peruvian bark in the cure of inter-
mittent fever. The berries, also, possess medicinal properties, but are rarely used. The bark furnishes bird-
lime, and the wood, which is white, and remarkably
close grained, is much used by cabinet-makers in inlay-
ing. *I. Paraguayensis* furnishes *Maté*, or Paraguay
Tea, which is so extensively used in Brazil and other
parts of South America ; for a full account of which, see
Forest Trees of Britain, vol. ii.

1. Ilex (Holly).—*Calyx* 4—5 cleft ; *corolla* wheel-
shaped, 4—5 cleft ; *stamens* 4—5 ; *stigmas* 4—5 ; *berry*
round, containing 4—5 bony seeds. (Name applied by
the Latins to some tree, though not our Holly.)

1. Ilex (*Holly*).

1. *I. Aquifolium* (Holly).—The only British species,
for a full account of which see *Forest Trees of Britain*,
vol. ii. The name *Aquifolium* means *needle-leaved*.
Holly is a corruption of the word "holy," from the use

to which its boughs are applied in ornamenting churches at Christmas.—Fl. May, June. Tree.

ILEX AQUIFOLIUM (*Holly*).

Ord. LIII.—OLEACEÆ.—The Olive Tribe.

Calyx divided, not falling off; *corolla* of 1 petal, 4-cleft, sometimes wanting ; *stamens* 2, alternate with

the lobes of the corolla ; *ovary* 2-celled ; *cells* 2-seeded ;
style 1 ; *fruit* a berry, drupe, or capsule, of 2 cells, each
cell often perfecting only a single seed.—Trees or shrubs,
the branches of which often end in conspicuous buds ;
the leaves are opposite, either simple or pinnate; the
flowers grow in clusters, or panicles.   The plants of this
Order inhabit the temperate regions of many parts of
the world.   By far the most important among them is
the plant from which the Order takes its name, *Olea*, the
Olive, among the earliest of plants cultivated by man.
The bark of the Olive is bitter and astringent, the wood
remarkably close grained and durable.   The fruit is a
drupe, or hard bony seed, enclosed in a fleshy, closely-
fitting case.   From this outer coat, and not from the
seed itself, oil is obtained by pressure.   Several kinds of
ash (*Fráxinus* and *Ornus*) produce manna, and are
valued for the strength and elasticity of their timber.
(For a more detailed account of the Ash, see *Forest Trees
of Britain.*)

1. LIGUSTRUM (Privet).—*Corolla* funnel-shaped, 4-
cleft ; *calyx* with 4 small teeth ; *fruit* a 2-celled berry.
(Name from the Latin name of the plant, and that from
ligo, to bind, from the use made of its twigs.)

2. FRÁXINUS (Ash).—*Calyx* 4-cleft, or 0 ; *corolla* 0 ;
fruit a winged 2-celled capsule. (Name, the Latin name
of the tree, denoting the ease with which it may be
split.)

1. LIGUSTRUM (*Privet*).

1. *L.vulgáré* (Privet).—The only British species.—
A common hedge-bush, with opposite, narrow-elliptical,
evergreen *leaves*, dense panicles of white, sickly-smelling
flowers, and black, shining *berries*, about the size of cur-
rants.   It is much used for hedges, especially in con-
junction with White-thorn, over which it has the advan-
tage of being a rapid grower.   It is commonly planted

to form divisions in town gardens, not being liable to be injured by smoke.—Fl. May, June.   Shrub.

LIGUSTRUM VULGARE (*Privet*).

2. FRÁXINUS (*Ash*).

1. *F. excelsior* (Ash).—*Calyx* and *corolla* both want-ing ; *leaves* pinnate, with an odd leaflet.—Woods and hedges ; common.   A noble tree, characterised by the light ash-coloured, smooth bark of its younger branches, of which the lower ones droop, and curve upwards again

at the extremities; by its large, black, terminal buds, the twigs supporting which are flattened at the end, and by its gracefully-feathered foliage. The tufts of seed-vessels, popularly called " keys," remain attached to the tree until the succeeding spring. A variety is occasionally found with undivided leaves, but it is not so handsome as the common form of the tree.—Fl. April and May, forming, at first, fruit-like, terminal heads, and, finally, loose panicles. Tree.

FRAXINUS EXCELSIOR (*Ash*).

Ord. LIV.—APOCYNEÆ.—Periwinkle Tribe.

Calyx deeply 5-cleft, not falling off ; *corolla* regular, 5-lobed, the lobes twisted when in bud, and when expanded having the sides of the margin unequally curved; *stamens* 5, inserted in the tube of the corolla; *anthers* distinctly 2-celled; *pollen* large; *ovary* 2-celled, or double; *pistil* resembling the shaft of a pillar with a double capital; *fruit* various.—Trees, shrubs, or herbaceous plants, with showy flowers, remarkable for the twisted lobes of the corolla when in bud, and yet more so for the symmetrical pistil. Many of them abound in a milky juice, and a large portion are poisonous. *Tanghinia venenifera* is the most deadly of known vegetable poisons, a single seed of which, though not larger than an almond, is sufficient to destroy twenty people. (For an account of the horrible use to which it was formerly applied in Madagascar, see *Wonders of the Vegetable Kingdom.*) The beautiful *Oleander*, (Nerium Oleander,) a common greenhouse shrub, is also a formidable poison, the powdered wood of which is used to destroy rats. In 1809, when the French troops were lying before Madrid, some of the soldiers went a marauding, every one bringing back such provisions as could be found. One soldier formed the unfortunate idea of cutting the branches of the Oleander for spits and skewers for the meat when roasting. This tree, it may be observed, is very common in Spain, where it attains considerable dimensions. The wood having been stripped of its bark, and brought in contact with the meat, was productive of most direful consequences; for, of twelve soldiers who ate of the roast, seven died, and the other five were dangerously ill. Some species, in which the characteristic properties are moderated, are, however, used as medicines. The tree called Hya-Hya by the natives of tropical America is one of the *Cow-trees,* which derive their name from the large quantity of sweet and wholesome

milk afforded by their branches. Several species furnish
caoutchouc, or India-rubber, of good quality. The only
British plants of this Order are the two Periwinkles,
which are astringent and acrid.

1. VINCA (Periwinkle).—*Corolla* salver-shaped, with
5 angles at the mouth of the tube, 5-lobed, the lobes
oblique ; *fruit* consisting of two erect horn-like capsules,
which do not burst. (Name from the Latin *vincio*, to
bind, from the cord-like stems.)

VINCA MINOR (*Lesser Periwinkle*).

VINCA (*Periwinkle*).

1. *V. major* (Greater Periwinkle).—*Stem* nearly erect;
leaves egg-shaped, with the margins minutely fringed.
—A doubtful native, being found only in the neighbourhood of dwelling-houses. A handsome plant, with
large, deep-green leaves, which are smooth, except at the

margins; and large, purplish-blue flowers, the mouth of which is angular, and the tube closed with hairs and the curiously curved anthers. The pistil of this flower, as well as of the following species, is a singularly beautiful object.—Fl. May—July. Perennial.

2. *V. minor* (Lesser Periwinkle).—*Stem* trailing, sending up short, erect, leafy shoots, which bear the flowers; margins of the *leaves* not fringed.—Woods, especially in the West of England, where it often entirely covers the ground with its evergreen leaves. It is much smaller than the last. A white variety occurs in Devonshire, and in gardens it is often met with bearing variegated leaves and double purple, blue, or white flowers.—Fl. March—June. Perennial.

Ord. LV.—GENTIANEÆ.—The Gentian Tribe.

Calyx usually 5, sometimes 4—8-cleft, not falling off; *corolla* of one petal, its lobes equalling in number those of the calyx, not falling off, twisted when in bud, often fringed about the mouth of the tube; *stamens* equalling in number the lobes of the corolla, and alternate with them; *ovary* of 2 carpels, 1- or imperfectly 2-celled; *style* 1; *stigmas* 2; *fruit* a many-seeded capsule or berry.—Mostly herbaceous plants, with opposite, generally sessile, leaves, and often large, brilliantly coloured flowers. This is an extensive Order, containing about 450 species, which are distributed throughout all climates, from the regions of perpetual snow to the hottest regions of South America and India. Though able to bear the most intense cold, they are very rare both in the Arctic and Antarctic regions. Under the equator, the lowest elevation at which they have been found is 7,852 feet: on the Himalaya range, one species has been found at a height of 16,000 feet; another in Ceylon at 8,000 feet: in southern Europe, one species, *Gentiana prostrata*, flourishes at between 6,000 and 9,000

feet: the same species occurs in the Rocky Mountains
of America at an elevation of 16,000 feet; and in the
Straits of Magellan and Behring's Straits, just above the
level of the sea. In South America and New Zealand,
the prevailing colour of the flower is red; in Europe,
blue; yellow and white being of rare occurrence. All
the known species are remarkable for the intensely bitter
properties residing in every part of the herbage, hence
they are valuable tonic medicines. That most commonly used in Europe is *Gentiana lutea* (Yellow Gentian); but there is little doubt that other species might
be employed with equally good effect.

1. GENTIANA (Gentian).—*Calyx* 4—5-cleft; *corolla*
funnel or salver shaped; *stamens* 5, rarely 4; *stigmas* 2.
(Name from *Gentius*, an ancient king of Illyria, who
discovered its medicinal value.)

2. ERYTHRÆA (Centaury).—*Calyx* 5-cleft; *corolla*
funnel-shaped, 5-cleft, not falling off; *stamens* 5; *anthers*
becoming spirally twisted; *stigmas* 2; *capsule* nearly
cylindrical, imperfectly 2-celled. (Name from the
Greek *erythros*, red, from the colour of the flowers.)

3. EXÁCUM (Gentianella).—*Calyx* 4-cleft, tubular;
corolla funnel-shaped, 4-cleft; *stamens* 4; *anthers* not
twisted; *stigma* undivided. (Name of Greek etymology,
and applied originally to some plant which was supposed
to have the power of *expelling* poison from the system.)

4. CHLORA (Yellow-wort). — *Calyx* deeply 8-cleft;
corolla with a very short tube, 8-cleft; *stamens* 8;
stigma 2—4-cleft. (Name from the Greek *chloros*,
yellow, from the colour of the flowers.)

5. MENYANTHES (Buck-bean).—*Calyx* deeply 5-cleft;
corolla funnel-shaped, with 5 lobes, fringed all over the
inner surface; *stamens* 5; *stigma* 2-lobed. (Name of
doubtful origin.)

6. VILLARSIA.—*Calyx* deeply 5-cleft, *corolla* wheel-
shaped, with 5 lobes, which are fringed only at the base;
stamens 5; *stigma* with 2 toothed lobes. (Name in
honour of M. de Villars, a French botanist.)

GENTIANA CAMPESTRIS (*Field Gentian*).

1. GENTIANA (*Gentian*).

1. *G. Amarella* (Autumnal Gentian).—*Stem* erect, branched, many-flowered ; *calyx* 5-cleft ; *corolla* salver-shaped, 5-cleft, fringed in the throat.—Dry chalky pastures, not common. A remarkably erect plant, with a

D

square, leafy, purplish stem, 6—12 inches high, and numerous, rather large, purplish blue flowers, which only expand in bright sunshine.—Fl. August, September. Annual.

2. *G. campestris* (Field Gentian).—*Stem* erect, branched, many-flowered ; *calyx* 4-cleft, the two outer lobes much larger ; *corolla* salver-shaped, 4-cleft, fringed in the throat.—Dry pastures, common. Resembling the last in habit, but at once distinguished by its 4-cleft flowers, which are of a dull purplish colour.—Fl. August, September. Annual.

3. *G. Pneumonanthé* (Marsh Gentian).—*Stem* erect, few-flowered ; *calyx* 5-cleft ; *corolla* between bell and funnel-shaped, 5-cleft, not fringed.—Boggy heaths, rare. Well distinguished from the preceding by its large, bell-shaped, deep blue flowers with 5 green stripes. There are rarely more than 1 or 2 flowers on the same stalk.— Fl. August, September. Annual.

* *G. verna* (Spring Gentian) is a very rare species, having on each stem a single large, intensely blue *flower*, which is 5-cleft, and has between the lobes 5 smaller 2-cleft segments : it has been found only in one or two places in Ireland, and at Teesdale, Durham. *G. nivalis* (Small Alpine Gentian) is also very rare, occurring only on the summits of the Highland mountains : the flowers grow several on a stem, and in colour and shape resemble the last.

2. ERYTHRÆA (*Centaury*).

1. *E. Centaurium* (Common Centaury).—A pretty herbaceous plant, 2—18 inches high, with square erect stems, which are much branched above, and terminate in variously divided flat tufts of small rose-coloured flowers ; the *leaves* are oblong, with strong parallel ribs, and remarkably smooth ; the *flowers* only expand in fine weather.—This is the common form of the plant as it occurs in dry fields and waste places. In other situations it varies so greatly, that some botanists enumerate

several distinct species, namely : *E. pulchella* (Dwarf Centaury), a minute plant, 2—8 inches high, with an exceedingly slender *stem*, and a few stalked *flowers* (often

ERYTHRÆA CENTAURIUM (*Common Centaury*).

only one); this is found on the sandy sea-shore, especially in the West of England : *E. littoralis* (Dwarf Tufted Centaury), a stunted plant, with broad *leaves*, and all the *flowers* crowded into a kind of head ; this occurs on turfy sea-cliffs : and *E. latifolia* (Broad-leaved Centaury), which has even broader *leaves* than the last, and bears its *flowers* in forked tufts, the main *stem* being divided into three branches. There are other minute differences, for which the student may consult more scientific works. The genus was formerly called *Chironia,* from the Centaur, *Chiron,* who was famous in

Greek mythology for his skill in medicinal herbs. The English name, Centaury, has the same origin.—Fl. July, August. Annual.

3. Exácum (*Gentianella.*)

1. *E. filiformé* (Least Gentianella).—The only British species. A minute, slender plant, in habit resembling *Erythræa pulchella*, and growing to about the same size, 2—4 inches; the *leaves* are very narrow, and soon wither; the *flowers* are yellow, and expand only in bright sunshine.—It grows in sandy heaths where water has stood during the winter. It is described by some authors under the name of *Cicendia filiformis*.—Fl. July. Annual.

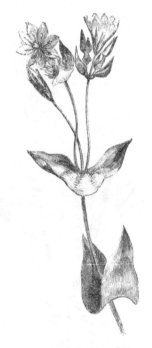

CHLORA PERFOLIATA (*Perfoliate Yellow-wort*).

4. CHLORA (*Yellow-wort.*)

1. *C. perfoliata* (Perfoliate Yellow-wort).—The only British species.—Chalk and limestone pastures; not uncommon. An erect plant, 12—18 inches high, remarkable for its glaucous hue, and for its pairs of *leaves*, which are rather distant, being united at the base (connate), with the stem passing through them; hence its name, *Perfoliate.* The flowers, which are large and handsome, are of a pale yellow, and expand only during sunshine. Fl. June, September. Annual.

MENYANTHES TRIFOLIATA (*Buck-bean, March Trefoil*).

5. MENYANTHES (*Buck-bean*).

1. *M. trifoliata* (Buck-bean, March Trefoil).—The only species; most common in spongy bogs, or in stagnant water. The only British plant belonging to the Order which has divided leaves. The *stem* scarcely rises above the soil or water in which it grows, but is over-

topped by the large ternate (composed of 3 leaflets)
leaves, which in shape and colour resemble those of the
Windsor Bean ; each leaf-stalk has a sheathing base,
opposite to one of which rises a compound cluster of
exceedingly beautiful flowers, which when in bud are of
a bright rose colour, and when fully expanded have the
inner surface of the corolla thickly covered with white
fringe. The root, which is intensely bitter, is said to
be the most valuable of known tonics. Fl. June, July.
Perennial.

6. VILLARSIA.

1. *V. nymphœoídes* (Water Villarsia). — The only
British species. A rare, floating aquatic, found in some
of the still ditches communicating with the Thames, and
in a few other places. As its specific name implies, it
has the habit of a Water-lily ; the leaves are nearly
round ; the flowers large, yellow, and fringed. Fl. July,
August. Perennial.

ORD. LVI.—POLEMONIACEÆ.—JACOB'S LADDER TRIBE.

Calyx deeply 5-cleft, not falling off ; *corolla* regular,
5-lobed ; *stamens* 5, from the middle of the tube of the
corolla ; *ovary* 3-celled ; *style* single ; *stigma* 3-cleft ;
capsule 3-celled, 3-valved.—Herbaceous plants, often
with showy flowers, which are remarkable for the blue
colour of their pollen. They are most common in the
temperate parts of America ; but within the tropics are
unknown. None of the species possess remarkable properties, but several are favourite garden flowers, as *Phlox,
Gilia, Polemonium,* and *Cobœa.*
1. POLEMONIUM (Jacob's Ladder). — *Corolla* wheel-
shaped, with erect lobes ; *stamens* bearded at the base ;
cells of the capsule many-seeded. (Name, the Greek
name of the plant.)

P. CŒRULEUM (*Greek Valerian, Blue Jacob's Ladder*).

1. POLEMONIUM (*Greek Valerian*).

1. *P. cœrúleum* (Greek Valerian, Blue Jacob's Ladder).
—The only British species, occasionally found in woods
in the north, but rare; a common garden flower, not
easily rooted out where it has once established itself.
A tall, erect plant, 1—2 feet high, with an angular
stem; pinnate, smooth *leaves;* and numerous terminal,
large, blue or white *flowers.*—Fl. June, July. Perennial.

Ord. LVII.—CONVOLVULACEÆ.—The Bindweed Tribe.

Calyx in 5 divisions, imbricated, often very unequal, not falling off; *corolla* of one petal, regular, plaited; *stamens* 5, from the base of the corolla; *ovary* 2—4 celled, few-seeded, surrounded below by a fleshy ring; *style* 1; *stigmas* 2; *capsule* 1—4 celled.—An extensive and highly valuable tribe of plants, most of which are herbaceous climbers, with large and very beautiful flowers. They are most abundant within the tropics, where they are among the most ornamental of climbing plants. As medicines, also, they occupy an important station. The roots of *Convólvulus Scammonia*, a Syrian species, furnishes scammony; jalap is prepared from a resin which abounds in the roots of several kinds of *Exogonium*, a beautiful climber, with long crimson flowers; and *Convólvulus Batatas* is no less valuable in tropical countries, supplying the sweet potato, the roots of which abound in starch and sugar, and are a nourishing food. *Cúscuta*, (Dodder,) is a parasitic genus, with branched, climbing, cord-like stems, no leaves, and globular heads of small wax-like flowers. The seeds germinate in the ground, and the young plants climb the stems of adjoining plants; and when they have taken root in them, lose their connexion with the ground. One British species is very abundant on the Furze; another on Flax, with the seeds of which it is supposed to be introduced; and a third grows on Thistles and Nettles.

1. Convólvulus (Bindweed).—*Corolla* trumpet-shaped, with 5 plaits and 5 very shallow lobes; *style* 1; *stigmas* 2. (Name, from the Latin, *convolvo*, to entwine, from the twisting habit of many species.)

2. Cúscuta (Dodder).—*Calyx* 4—5 cleft; *corolla* bell-shaped, 4—5 cleft, with 4—5 scales at the base within. (Name, said to be derived from the Arabic, *Kechout.*)

1. Convólvulus (*Bind-weed*).

1. *C. arvensis* (Field Bind-weed).—*Stem* climbing ;
leaves arrow-shaped, with acute lobes ; *flowers* 1—3
together ; *bracts* minute, distant from the flower.—A
common weed in light soil, either trailing along the
ground among short grass, or climbing wherever it finds
a support.   The flowers are rose-coloured, with dark
plaits, handsome and fragrant, opening only in sunny
weather.—Fl. June, July.   Perennial.

2. *C. sépium* (Great Bind-weed).—*Stem* climbing ;
leaves arrow-shaped, with abrupt lobes ; *flowers* solitary,
on square stalks ; *bracts* large, heart-shaped, close to the

CONVOLVULUS SOLDANELLA (*Sea Bind-weed.*)

flower.—In bushy places, common, and a most mischievous weed in gardens, not only exhausting the soil
with its roots, but strangling with its twining stems the
plants which grow near. The flowers are among the
largest which this country produces ; while in bud they
are entirely enclosed in the large bracts, and when
expanded are pure white and very handsome. The
fruit is not often perfected.—Fl. July—September.
Perennial.

3. *C. Soldanella* (Sea Bind-weed).—*Stem* not climbing ; *leaves* fleshy, roundish, or kidney-shaped ; *flowers*
solitary, on 4-sided, winged stalks ; *bracts* large, egg-
shaped, close to the flower.—A very beautiful species,
growing only on the sandy sea-coast, and decorating the
sloping sides of sand-hills with its large, pale, rose-
coloured flowers, striped with red. The stems are usually
almost entirely buried beneath the sand, and the flowers
and leaves merely rise above the surface. The flowers,
which are nearly as large as those of the preceding
species, expand in the morning, and in bright weather
close before night.—Fl. June—August. Perennial.

2. CÚSCUTA (*Dodder*).

1. *C. Epithymum* (Lesser Dodder).—*Stems* parasitical,
thread-like, branched ; *flowers* in dense, sessile heads ;
tube of the *corolla* longer than the calyx ; *style* longer
than the corolla.—Parasitic on Heath, Thyme, Milk vetch,
Potentilla, and other small plants, but most abundant on
Furze, which it often entirely conceals with tangled masses
of red, thread-like stems. The flowers are small, light
flesh-coloured, and wax-like. Soon after flowering the
stems turn dark-brown, and in winter disappear.—Fl.
August, September. Annual.

* Other species of Dodder, which more or less resemble
the preceding species, are *C. Europæa* (Greater Dodder),
which is parasitical on Thistles and Nettles ; *C.Epilinum*
(Flax Dodder), parasitical on Flax, to crops of which it
is sometimes very destructive ; and *C. Trifolii* (Clover

Dodder), parasitical on Clover, with the seeds of which it is supposed to have been introduced.

CUSCUTA EPITHYMUM (*Lesser Dodder*).

Ord. LVIII.—BORAGINEÆ.—The Borage Tribe.

Calyx in 5, rarely 4, deep divisions, not falling off; *corolla* of one petal, 5 or rarely 4-cleft, frequently having valves or teeth at the mouth of the tube; *sta-*

mens 5 inserted into the corolla and alternate with its lobes ; *ovary* 4-parted, 4-seeded; *style* 1, rising from the base of the divided ovary; *fruit* consisting of 4, rarely 2, nut-like, distinct seeds, each enclosed in a pericarp.— Herbs or shrubs with alternate leaves, which are usually covered with hairs or bristles rising from a swollen base. This character was considered by Linnæus sufficiently constant to give to the natural order the name of ASPERIFOLIÆ, or Rough-leaved plants, but the present name of the order is now preferred as being more comprehensive, a few plants in it having perfectly smooth leaves. The Borage Tribe are natives principally of the temperate regions of the northern hemisphere, especially of the warmer parts, and are more numerous in the Old than the New World. Most of them bear their flowers in spikes or clusters, which are rolled up round the terminal flowers as a centre, and expand a few at a time. The prevailing colour is blue or purple, but many, when first opening, are of a reddish hue which subsequently deepens, so that it is not unusual to see flowers of different tints in the same spike or cluster. They possess no remarkable properties, but abound in a soft mucilaginous juice, which gives a coolness to beverages in which they are steeped, on which account Borage is a constant ingredient in the drink known as " cool tankard." The roots of Alkanet and some others contain a red substance, which is used as a dye. Comfrey (*Symphytum officinále*) is sometimes grown as an esculent vegetable, but is little valued except as food for horses. The plants of the genus *Myosótis* are popularly known by the name " Forget-me-not ; " the real Forget-me-not is *M. palustris*.

1. ECHIUM (Viper's Bugloss).—*Corolla* irregular, with an open mouth ; *stamens* unequal in length. (Name from the Greek, *echis*, a viper, against the bite of which it was formerly considered an antidote.)

2. PULMONARIA (Lungwort).—*Calyx* tubular, 5-cleft ; *corolla* funnel-shaped, its throat naked ; *stamens* enclosed

within the corolla. (Name from the Latin, *pulmo*, the lungs, which the spotted leaves were supposed to resemble.)

3. LITHOSPERMUM (Gromwell).—*Calyx* deeply 5-cleft; *corolla* funnel-shaped, its throat naked, or with 5 minute scales; *seeds* stony. (Name from the Greek, *lithos*, a stone, and *sperma*, seed, from the hardness of the seeds.)

4. SÝMPHYTUM (Comfrey).—*Calyx* deeply 5-cleft; *corolla* bell-shaped, closed with 5 awl-shaped scales. (Name from the Greek *symphyo*, to unite, from its imagined healing qualities.)

5. BORÁGO (Borage).—*Calyx* deeply 5-cleft; *corolla* wheel-shaped, its throat closed with 5 short, erect, notched scales; *stamens* forked. (Name, a corruption of *corago*, from *cor*, the heart, and *ago*, to bring, from its use in stimulating drinks.)

6. LYCOPSIS (Bugloss).—*Calyx* deeply 5-cleft; *corolla* funnel-shaped, with a bent tube, its throat closed by prominent blunt scales. (Name in Greek signifying " a wolf's face," from some fancied resemblance between the flower and a wolf's head.)

7. ANCHÚSA (Alkanet).—*Calyx* deeply 5-cleft; *corolla* funnel or salver-shaped, with a straight tube, its throat closed by prominent blunt scales. (Name from the Greek, *anchousa*, paint, from the use of its roots as a dye.)

8. MYOSÓTIS (Scorpion Grass, Forget-me-not).—*Calyx* 5-cleft; *corolla* salver-shaped, its lobes blunt, twisted when in bud, and its throat nearly closed by blunt scales. (Name in Greek signifying a *mouse's ear*, from the shape of the leaves.)

9. ASPERÚGO (Madwort).—*Calyx* 5-cleft, with alternate smaller teeth; *corolla* funnel-shaped, with rounded scales in the throat. (Name from the Latin *asper*, rough, from the excessive roughness of the leaves.)

10. CYNOGLOSSUM (Hound's-tongue).—*Calyx* 5-cleft; *corolla* funnel-shaped, with a short tube, its mouth closed by prominent blunt scales; *nuts* flattened, prickly. (Name in Greek signifying a *dog's tongue*, from the shape and size of the leaves.)

ECHIUM VULGARE (*Common Viper's Bugloss*).

1. ÉCHIUM (*Viper's Bugloss*).

1. *E. Vulgáre* (Common Viper's Bugloss).—Rough, with prickly bristles; *leaves* narrow, tapering; *flowers* in short lateral spikes; *stamens* longer than the corolla.— Walls, old quarries and gravel-pits. A handsome plant 2—3 feet high, remarkable for its bristly or almost prickly stems and leaves, and numerous curved spikes of flowers, which on their first opening are bright rose-

coloured and finally of a brilliant blue. The roots are very long, and descend perpendicularly into the loose soil in which the plant usually grows. A variety is occasionally found with white flowers. The name Bugloss, which is of Greek origin, signifies an *ox's tongue*, from the roughness and shape of the leaves.—Fl. June, July. Biennial.

PULMONARIA OFFICINALIS (*Common Lungwort*).

2. Pulmonaria (*Lungwort*).

1. *P. officinalis* (Common Lungwort).—*Leaves* rough, lower ones stalked, egg-shaped or heart-shaped at the base, upper, egg-shaped sessile.—Woods and thickets, rare; often an outcast from gardens, where it is cultivated for the sake of its ornamental leaves, which are curiously spotted with white, and for its purple flowers. Fl. May. Perennial.

* *P. angustifolia* has narrower leaves than the last, and is a taller plant; it occurs in Hampshire and the Isle of Wight, but is far from common.

3. Lithospermum (*Gromwell*).

1. *L. officinálé* (Common Gromwell, or Grey Millet).— *Stem* erect, much branched towards the summit; *leaves* oblong, tapering to a point, bristly above, hairy beneath; *nuts* polished.—Dry, bushy, and stony places; not unfrequent. Distinguished by its erect stems, 2 feet high, which generally grow 5 or 6 from the same root, by its small yellowish white flowers, and above all, by its highly polished seeds, which resemble small sea-shells, and contain a considerable quantity of pure silica or flint.—Fl. June, July. Perennial.

2. *L. arvensé* (Corn Gromwell). — *Stem* branched; *leaves* narrow, tapering, hairy; *nuts* wrinkled.—Cornfields; less common than the preceding, but not rare. Stem about a foot high, branched from the lower part, and having rather small white flowers, the calyx of which lengthens when in fruit, and contains 3 or 4 brown, wrinkled seeds.—Fl. May—July. Annual.

* *L. marítimum* (Sea-side Gromwell) is a singular plant inhabiting the sea-coast of North Wales, Scotland and Ireland; the *leaves* are fleshy, and covered with a glaucous bloom; they are destitute of bristles, but are

sprinkled with hard dots which are very evident in
dried specimens; the plant when fresh is said to have
the flavour of oysters; the *flowers* are purplish blue:
L. púrpuro-cœrúleum (Purple Gromwell) is a rare species growing in a chalky or limestone soil, and is distinguished by its prostrate barren *stems*, and large
bright blue *flowers*.

LITHOSPERMUM OFFICINALE
(Common Gromwell, or Grey Millet)

4. SÝMPHYTUM (*Comfrey*).

1. *S. officinále* (Common Comfrey).—*Stem* winged in
the upper part; *leaves* elliptical, pointed, tapering
towards the base and running down the stem; *flowers*
drooping, in 2-forked clusters.—Watery places and banks

of rivers, common. A large and handsome plant 2—3 feet high, with branched leafy stems, and several clusters of white, pink, or purple drooping flowers. Often introduced into gardens, from which it is very difficult to eradicate it when it has once established itself, owing to the brittleness of its fleshy roots, the least bit of which will grow.—Fl. May—August. Perennial.

SYMPHYTUM OFFICINALE (*Common Comfrey*).

* *S. tuberosum* (Tuberous Comfrey) is a more slender plant than the preceding, and is very rare, except in Scotland ; the stem is scarcely branched, and but slightly winged.

BORAGO OFFICINALIS (*Common Borage*).

5. Borágo (*Borage*).

1. *B. officinális* (Common Borage).—The only British species, not unfrequent in waste ground, especially near buildings. The stems are about 2 feet high, and, as well as the leaves, are thickly covered with stiff whitish bristles; the flowers, which are large, deep blue, and very handsome, grow in terminal drooping clusters, and may readily be distinguished from any other plant in the order by their prominent black anthers. The juice has the smell and flavour of cucumber, and is an ingredient in the beverage called "cool tankard." A variety some-

times occurs with white flowers.—Fl. June—September. Biennial.

LYCOPSIS ARVENSIS (*Small Bugloss*).

6. Lycopsis (*Bugloss*).

1. *L. arvensis* (Small Bugloss).—The only British species, common in waste ground, and on the sea-coast. A branched prickly plant 6—18 inches high, with oblong wavy *leaves*, and forked, curved clusters of minute blue *flowers*, the tube of which is bent, and in this respect is unlike any other British plants of the order.—Fl. June—August. Annual.

ANCHUSA SEMPERVIRENS (*Evergreen Alkanet*).

7. ANCHÚSA (*Alkanet*).

1. *A. sempervírens* (Evergreen Alkanet).—A stout
bristly plant, with deep green, egg-shaped *leaves*, and
short spikes of rather large, salver-shaped *flowers*, which
are of an intense azure blue. It is not generally considered to be a native, but in Devonshire it is not an
uncommon hedge-plant.—Fl. June—August. Perennial.

* *A. officinális* (Common Alkanet) has purple, funnel-shaped *flowers*, which grow in 1-sided spikes, the
segments of the *calyx* being longer than the *corolla*.
It is frequent in gardens, from which it is supposed
to have escaped; but it is not common in a wild state.

8. Myosótis (*Mouse-ear, Scorpion-grass, Forget-me-not*).

1. *M. palustris* (Forget-me-not).—*Calyx* covered with straight, closely-pressed bristles, open when in fruit; *root* creeping.—Watery places, common. Few plants have been more written about than the Forget-me-not, yet there is great disagreement among writers as to the plant to which the name properly belongs. Some appear to have had the *Alkanet* in view; others, the *Speedwell;* and others, again, some of the smaller species of *Myosótis*, which last, though very like the true Forget-me-not, are inferior in size and brilliancy of colour. The real Forget-me-not is an aquatic plant, with a long rooting stem, bright green, roughish leaves, and terminal, leafless, 1 sided clusters of bright blue flowers, with a yellow eye, and a small white ray at the base of each lobe of the corolla. The species which is most like it is *M. repens* (Creeping Water Scorpion-grass), which, as its name implies, has also a creeping root; the hairs of the calyx are closely pressed, as in *M. palustris*, but the calyx is closed when in fruit, and the clusters of flowers usually have a few leaves on the stalk. *M. cæspitosa* (Tufted Water Scorpion-grass), resembles the above, but has a fibrous root, and the flowers of both the last are smaller than those of *M. palustris*. All three grow in watery places; *M. palustris* being most common, and flowering from June to October, *M. repens* least so, and, as well as *M. cæspitosa*, not flowering so late in the year. Three other and yet smaller species are common, but these do not grow in watery places, and are of a different habit.

2. *M. arvensis* (Field Scorpion-grass).—*Calyx* covered with spreading, hooked bristles, closed when in fruit; *stalks* of the fruit spreading.—In cultivated ground, hedges, and groves, abundant. Whole plant roughish, with spreading bristles; the stems are from 6—18 inches high, or more; the flowers blue, small, but very beau-

tiful.   This is the commonest species of all.—Fl. June—
August.   Annual.

3. *M. collina* (Early Field Scorpion-grass).—*Calyx*
covered with spreading hooked bristles, open when in
fruit ; *cluster* with a solitary flower in the axil of the
uppermost leaf.—Dry banks, not uncommon, but frequently overlooked in consequence of its minute size.
The whole plant rarely exceeds 3 inches in length ; the
stems usually spread near the ground, and terminate in
clusters of very minute bright blue flowers,(never pink or
yellow). On its first appearance, in April, the flowers are

buried among the leaves, but the stems finally lengthen into clusters, and as the season advances the whole plant dries up, and disappears.—Fl. April, May. Annual.

4. *M. versicolor* (Party-coloured Scorpion-grass).— *Calyx* covered with spreading, hooked bristles, closed when in fruit ; *cluster* on a long leafless stalk ; *stalk* of the fruit erect.—Fields and banks, common. A very distinct species, 3—6 inches high ; the stem is leafy below, naked above, and ends in a cluster of flowers, which are singularly coiled up when in bud, and when they expand, are either blue or yellow on the same plant. —Fl. April—June. Annual.

MYOSOTIS ALPESTRIS (*Mountain Forget-me-not*).

* There are two other British species of *Myosótis*, which approach the Forget-me-not in the size and beauty of their flowers—*M. alpestris*, which grows on the Scottish mountains, and *M. sylvática*, which is almost confined to Scotland and the north of England, where it grows in shady places. Both these species have spreading bristles on the calyx.

CYNOGLOSSUM OFFICINALE (*Common Hound's-Tongue*)

9. ASPERÚGO (*Madwort*).

1. *A. procumbens* (German Madwort).—The only species occurring, very sparingly, in Scotland and the north of England. The *stems* are prostrate, angular, and thickly set with rigid, curved bristles ; the *flowers* are small, blue, and solitary in the axils of the upper leaves.—Fl. June, July. Annual.

10. CYNOGLOSSUM (*Hound's-tongue*).

1. *C. officinále* (Common Hound's-tongue). —*Leaves* downy.—Waste ground, especially near the sea. A stout herbaceous plant, 1—2 feet high, with large downy leaves, lurid-purple flowers, and large flattened seeds, which are covered with barbed prickles, and stick to the wool of animals or the clothes of passengers as closely as burs. The whole plant has a strong disagreeable smell, like that of mice.—Fl. June—August. Biennial.

* *C. sylváticum* (Green-leaved Hound's-tongue) is a plant of very local occurrence ; the *leaves* are shining above (not downy), and the *flowers* reddish, changing to blue.

ORD. LIX.—SOLANEÆ.—NIGHTSHADE TRIBE.

Calyx deeply 5- rarely 4-cleft, inferior; *corolla* of one *petal*, 5- or rarely 4-cleft, equal or nearly so, plaited when in bud ; *stamens* equalling in number the divisions of the corolla, and alternate with them ; *anthers* bursting lengthwise, or opening by pores ; *ovary* 2-celled ; *style* 1 ; *stigma* simple ; *fruit* a 2- or 4-celled capsule or berry ; *seeds* numerous.—A large and highly important order, containing about 900 species of herbaceous plants or shrubs, which inhabit most parts of the world except the coldest, and are most abundant within the tropics. The prevailing property of plants belonging to the Nightshade Tribe is narcotic, and many are, in consequence, highly poisonous ; in others, certain parts of the plant have poisonous properties, the rest being harmless, and some even contain a large quantity of nutritious matter. The genus *Solánum* is a very extensive one, comprising as many as 600 species. First among these in importance stands the Potato (*S. tuberósum*), a native of Chili, Lima, Quito, and Mexico, which was introduced into Spain in the early part of the 16th century, and into Ireland by the colonists sent out by Sir Walter Raleigh, who brought it from Virginia in 1586. It was first planted on Sir Walter Raleigh's estate at Youghall, near Cork, and was cultivated for food in that country long before its value was known in England. Its leaves and berries are narcotic, but its tubers contain no noxious matter, abounding in an almost tasteless starch; on which account it is less liable to cloy on the palate than any other vegetable food except bread. *S. Melóngena* (the Egg-plant), a common greenhouse plant, is remarkable for bearing a large berry of the size and colour of a poulet's egg. *S. Dulcamára*, (Nightshade, or Bittersweet,) a common English plant, with purple and yellow flowers, has narcotic leaves and scarlet berries, which possess the same property. *S. nigrum*, a smaller species, a common weed in England and most other countries, except the coldest, has white flowers and black berries. It is narcotic to a dangerous degree. *Atropa Belladonna*, a stout herbaceous plant, with dingy purple bell-shaped flowers, is the Deadly Nightshade, so called from the poisonous nature of every part of the plant, especially the berries, which are large, black and shining, and of a very attractive appearance. Its juice possesses the singular property of dilating the pupil of the eye, on which account it is extensively used by oculists when operations are to be performed. The Mandrake (*Mandrágora officinalis*) was anciently thought to possess miraculous properties. It was said to shriek when taken from the ground, and to cause the instant death of any one who heard its cries. The person who gathered it, therefore, always stopped his ears with cotton, and harnessed a dog to the root, who, in his efforts to escape, uprooted the plant, and instantly fell dead. The forked root was then trimmed so as to resemble the human form, a berry being left to represent the head. The fruit is eatable. Tobacco is the foliage of several species of *Nicotiana*, a violent poison when received into the stomach, though commonly employed in other ways without apparent ill effects. *Hyoscyamus niger*, or Henbane, is a stout herbaceous plant, with sticky, fœtid leaves, and cream-coloured flowers veined with purple ; it is a powerful narcotic, and in skilful hands scarcely less valuable than opium. *Datúra Stramonium* (Thorn-apple) bears large, white, trumpet-shaped flowers, and prickly seed-vessels ; it is also a dangerous poison, though employed with good effect in several nervous and other disorders, especially asthma. *Phýsalis Alkekengi* is the Winter Cherry, remarkable for bearing a crimson berry enclosed in the enlarged calyx, which, after exposure to the wet, decays, leaving the berry hanging within a case of net-work. The genus *Cápsicum* affords Cayenne pepper, which is prepared by grinding the dried seed-vessels with their contents. Finally, *Lycopérsicum* produces Tomatoes or Love Apples, which are much used in making sauce.

1. SOLÁNUM (Nightshade).—*Corolla* wheel-shaped, 5-cleft, the segments spreading or reflexed ; *anthers* opening by 2 pores at the summit ; *berry* roundish, with 2 or more cells. (Name of doubtful origin).

2. ÁTROPA (Deadly Nightshade).—*Corolla* bell-shaped, with 5 equal lobes ; *stamens* distant ; *berry* of two cells. (Name from *Átropos*, one of the Fates, who was supposed to cut the thread of human destiny.)

3. *Hyoscýamus* (Henbane).—*Corolla* funnel-shaped, with 5 unequal lobes ; *capsule* 2-celled, closed by a lid. (Name in Greek signifying *Hog's bean*.)

1. SOLÁNUM (*Nightshade*).

1. *S. Dulcamára* (Woody Nightshade, Bittersweet).— *Stem* shrubby, climbing ; *leaves* heart-shaped, the upper ones eared at the base ; *flowers* drooping.—Damp hedges

and thickets, common. This plant, which is frequently, though incorrectly, called *Deadly Nightshade,* is well marked by its straggling, woody stem, which climbs among bushes to the length of 8 or 10 feet, and its purple flowers, the yellow anthers of which unite in the form of a cone. At the base of each lobe of the corolla are 2 green spots. The flowers grow in drooping, loose tufts, and are succeeded by scarlet berries, the length of which slightly exceeds the breadth.—Fl. June, July. Perennial.

SOLANUM DULCAMARA (*Woody Nightshade, Bittersweet*).

2. *S. nigrum* (Black Nightshade).—*Stem* herbaceous; *leaves* egg-shaped, wavy at the edge, and bluntly toothed; *flowers* drooping.—Waste ground, common. A branching herb, with drooping umbels of white flowers and black globular berries.—Fl. July—September. Annual.

ATROPA BELLADONNA (*Deadly Nightshade, Dwale*).

2. ÁTROPA (*Deadly Nightshade*).

1. *A. Belladonna* (Deadly Nightshade, Dwale).—A stout herbaceous plant, 3—4 feet high, with large egg-

shaped *leaves*, and solitary, drooping, bell-shaped *flowers*, which grow in the axils of the upper leaves, and are of a lurid-purple hue.   The berries are black, and as large as cherries, which they somewhat resemble in appearance, but may be readily distinguished by the calyx at the base.   This noxious plant, which is the most dangerous growing in Britain, on account of its active poisonous properties and the attractive appearance of its berries, is fortunately of rare occurrence, growing principally in old quarries and among ruins.   Buchanan relates that the Scots mixed the juice of Belladonna with the bread and drink, with which by their truce they were to supply the Danes, which so intoxicated them, that the Scots killed the greater part of Sweno's army while asleep.   The "insane root that takes the reason prisoner," mentioned by Shakspeare, is also thought to be this.   The English name *Dwale* is derived from a French word *deuil*, which signifies ' mourning.'—Fl. June—August.   Perennial.

3. HYOSCÝAMUS (*Henbane*).

1. *H. niger* (Common Henbane).—The only British species, common in waste places, especially on a chalky soil or near the sea.   An erect, branched, herbaceous plant, 2—3 feet high, with large, viscid, hairy *leaves*, and numerous funnel-shaped, cream-coloured *flowers* with purple veins and a dark eye.   The flowers are arranged in rows along one side of the stem, and are succeeded by 2-celled capsules, which are enclosed by the calyx, and covered by a lid which falls off when the seeds are ripe. The whole plant has an exceedingly disagreeable smell, and is dangerously narcotic, especially at the time when the seeds are ripening.   An extract is used in medicine, and is often of great service, producing the effect of opium, without the unpleasant symptoms which frequently follow the administration of that drug.   The capsules and seeds of Henbane, smoked like tobacco, are a rustic remedy for the tooth-ache; but convul-

HYOSCYAMUS NIGER (*Common Henbane*).

sions and temporary insanity are said to be sometimes the consequences of their use.—Fl. June, July. Annual or Biennial.

Ord. LX.—OROBANCHEÆ.—Broom-rape Tribe.

Calyx variously divided, not falling off ; *corolla* irregular, usually 2-lipped, imbricated in the bud ; *stamens*
4, 2 long and 2 short ; *anthers* often pointed or bearded
at the base ; *ovary* in a fleshy disk, many-seeded ; *style*
1 ; *stigma* 2-lobed ; *capsule* 2-valved ; *seeds* small,
numerous, attached to the valves of the capsule in 2—4
rows.—A tribe of herbaceous plants, distinguished by a
stout succulent stem, which is of a peculiar dingy red
hue, bearing no leaves, but more or less clothed with
taper-pointed scales, which are most abundant about the
swollen base of the stem.   The flowers are large for the
size of the plant, and in all British species are of nearly
the same hue as the stem, and arranged in a spike not
unlike a head of asparagus, with one or more scale-like
bracts at the base of each flower.   All the species are
parasitical on the roots of other plants.   The seeds, it is
said, will lie buried for some years in the ground without
vegetating, until they come in contact with the young
roots of some plant adapted to their wants, when they
immediately sprout, and seize on the points of the roots,
which swell, and serve as a base to the parasite.   There
are but two British genera belonging to this order,
Orobánché and *Lathrœa*, of which some attach themselves
to particular species ; others infest particular tribes, and
others, again, have a wider range of subjects.   Several
of those belonging to the genus *Orobánché* are very
difficult of discrimination ; botanists, indeed, are not
agreed as to the number of species ; some uniting under
a common name specimens found growing on various
plants ; others considering a slight variation in structure,
joined to a difference of situation, enough to constitute
a specific distinction.

1. Orobánché (Broom-rape).—*Calyx* of two lateral
sepals, which are usually 2-cleft, and often combined in
front, with 1—3 bracts at the base ; *corolla* gaping,

4—5 cleft, not falling off.—(Name from the Greek, *órobos*, a vetch, and *ancho*, to strangle, from the injurious effects produced in the plants to which they attach themselves.)

2. LATHRÆA (Tooth-wort).—*Calyx* bell-shaped, 4-cleft; *corolla* gaping, 2-lipped, the upper lip arched, entire, not falling off.—(Name in Greek signifying *concealed*, from the humble growth of the plants among dead leaves.)

1. OROBÁNCHÉ (*Broom-rape*).

* *Bracts one to each flower.*

1. *O. major* (Greater Broom-rape).—*Corolla* tubular, the lower lip in 3 lobes, of which the middle one is blunt, and longer than the others; *stamens* smooth below, downy above; *style* downy.—On the roots of Furze, Broom, and other plants of the order *Leguminosæ*, frequent. A stout, leafless, club-like plant, much swollen at the base, of a reddish-brown hue, viscid and clothed with tapering scales, which pass into bracts as they ascend the stem. The flowers are of a pinkish-brown hue, and are crowded into a dense spike. The juice is bitter and astringent, and has been used medicinally.—Fl. June, July. Perennial.

2. *O. minor* (Lesser Broom-rape).—*Stamens* hairy below, smooth above; *style* nearly smooth.—Under this description are included several species or varieties which are parasitical severally on Clover, Ivy, and Sea-Carrot. They all resemble the last in habit, but are of smaller size.

* To this group belong *O. caryophylládcea* (Clove-scented Broom-rape), a species with hairy *stamens*, and a dark purple *stigma*; growing in Kent, on the roots of *Galium Mollúgo*: *O. elatior*, a rare species, parasitical on *Centauréa scabiosa*: and *O. rubra*, abundant on basaltic rock in Scotland and the north of Ireland, and on magnesian rock at the Lizard Point, Cornwall. This

species appears to be parasitical on the roots of Wild
Thyme.

OROBANCHE MAJOR (*Greater Broom-rape*).

** *Bracts 3 under each flower.*

3. *O. ramósa* (Branched Broom-rape).—*Stem* branched.
—On the roots of Hemp, very rare. Distinguished
from the preceding by its lighter colour, and branched
stem.—Fl. August, September. Annual.

LATHRÆA SQUAMARIA (*Tooth-wort*).

* *O. cœrulea* is another rare species, found in Norfolk, Hertfordshire, and the Isle of Wight. It may be distinguished by its 3 bracts, and its bluish-purple hue.

2. LATHRÆA (*Tooth-wort*).

1. *L. squamaria* (Tooth-wort).—The only British species, growing, in woods and thickets, on the roots of the Hazel. The *stem* is branched below the surface of the ground, or withered leaves among which it grows ; it is of a lightish hue, and thickly clothed with tooth-like *scales ;* each *branch* bears a 1-sided cluster of drooping purplish *flowers*, with rather broad *bracts* at the base of each.—Fl. April, May. Perennial.

ORD. LXI —SCROPHULARINEÆ.—FIG-WORT TRIBE.

Calyx 4 – 5-lobed, not falling off; *corolla* irregular, often 2-lipped ; *stamens* usually 4, 2 long and 2 short, (didynamous,) sometimes 2 or 5 ; *ovary* 2-celled ; *style* 1; *stigma* 2-lobed ; *capsule* 2-celled, 2—4-valved, or opening by pores.—A large and important order, containing nearly two thousand species, of which some are shrubs, but the greater number are herbaceous, inhabiting all parts of the world, from the Arctic regions to the Tropics. The general character of the species is acrid and bitterish, and some have powerful medicinal properties. The powdered leaves of Foxglove (*Digitális purpúrea*) lower the pulse, and, if taken in large doses, are poisonous. *Euphrasia* (Eye-bright), the "Euphrasy" of Milton, makes a useful eye-water. Among foreign species, *Gratiola* is said to be the active ingredient in the famous gout medicine, "Eau médicinale." Foxglove, Snapdragon, Mullein, and Toad-flax, have showy and ornamental flowers ; and several kinds of Speedwell (*Verónica*) are deservedly admired for their small but elegant blue flowers.

* *Stamens* 4, 2 *long and* 2 *short* (*didynamous*).

1. Digitális (Foxglove).—*Calyx* in 5 deep, unequal segments; *corolla* irregularly bell-shaped, with 4—5 shallow lobes; *capsule* egg-shaped.—(Name from the Latin *digitál*é, the finger of a glove, which its flowers resemble.)

2. Antirrhínum (Snapdragon).—*Calyx* 5 parted; *corolla* personate, swollen at the base, (not spurred,) its mouth closed by a palate; *capsule* oblique, opening by pores at the top.—(Name in Greek signifying *opposite the nose*, from the mask-like appearance of the flowers.)

3. Linária (Toad-flax).—Like *Antirrhínum*, except that the *corolla* is spurred at the base.—(Name from *Linum*, Flax, which the leaves of some species resemble.)

4. Scrophularia (Fig-wort).—*Calyx* 5-lobed; *corolla* nearly globose, with 2 short lips, the upper 2-lobed, with a small scale within, the lower 3-lobed; *capsule* opening with 2 valves, the edges of which are turned in.— (Name from the disease for which the plant was formerly thought a specific.)

5. Limosella (Mudwort).—*Calyx* 5-cleft; *corolla* bell-shaped, 5-cleft, equal; *capsule* globose, 2-valved.— (Name from the Latin *limus*, mud, from the character of the places in which the plant grows.)

6. Melampýrum (Cow-wheat).—*Calyx* tubular, with 4 narrow teeth; *corolla* gaping, *upper lip* flattened vertically, turned back at the margin, *lower lip* 3-cleft; *capsule* oblong, obliquely pointed, flattened; *seeds* 1 or 2 in each cell.—(Name in Greek signifying *black wheat*, the seeds, when ground and mixed with flour, being said to make it black.)

7. Pediculáris (Red-rattle). — *Calyx* inflated, its segments somewhat leafy; *corolla* gaping, *upper lip* arched, flattened vertically, *lower lip* plane, 3-lobed; *capsule* flattened, oblique; *seeds* angular.—(Name in allusion to the disease produced in sheep which feed in places where it grows.)

8. RHINANTHUS (Yellow-rattle).—*Calyx* inflated, 4-toothed ; *corolla* gaping, *upper lip* flattened vertically, *lower lip* plane, 3-lobed ; *capsule* flattened, blunt ; *seeds* numerous, flat and bordered.—(Name in Greek signifying *nose-flower*, from its peculiar shape.)

9. BARTSIA.—*Calyx* tubular, 4-cleft ; *corolla* gaping, with a contracted throat, *upper lip* arched, entire, *lower lip* 3-lobed, lobes bent back ; *capsule* flattened, pointed ; *seeds* numerous, angular.—(Name in honour of John Bartsch, a Prussian botanist.)

10. EUPHRASIA (Eye-bright).—*Calyx* tubular, 4-cleft ; *corolla* gaping, *upper lip* divided, *lower lip* in 3 nearly equal lobes ; *anthers* spurred at the base ; *capsule* flattened, blunt or notched ; *seeds* numerous, ribbed — (Name from the Greek, *Euphrósyne*, gladness, from the valuable properties attributed to it.)

11. SIBTHORPIA (Cornish Money-wort).—*Calyx* in 5 deep spreading segments ; *corolla* wheel-shaped, 5-cleft, nearly regular ; *capsule* nearly round, flattened at the top.—(Name in honour of Dr. Sibthorp, formerly professor of botany at Oxford.)

** *Stamens* 2.

12. VERÓNICA (Speedwell).—*Corolla* wheel-shaped, unequally 4-cleft, lower segment the narrowest.—(*Verónica* is the name of a saint in the Romish Church, but why given to this plant is unknown.)

*** *Stamens* 5.

13. VERBASCUM (Mullein).—*Calyx* 5-parted ; *corolla* wheel-shaped, 5-cleft, irregular ; *stamens* hairy. — (Name from the Latin *barba*, a beard, from the shaggy leaves of some species.)

1. DIGITALIS (*Foxglove*).

1. *D. purpúrea* (Purple Foxglove).—The only British species, common in many of the hilly districts of Great

Britain, but almost unknown in the plains.   A stately
plant, 2—6 feet high, with large wrinkled *leaves,* and a
tall *stem,* bearing numerous handsome, purple, bell-

DIGITALIS PURPUREA (*Purple Foxglove*).

shaped *flowers,* which are arranged in the form of a
tapering *spike,* and droop after expansion.   On the inside,
the flowers are beautifully spotted ; occasionally they

are found of a pure white; but though this variety is elegant, it is by no means so striking a plant as the other. The name *Foxglove* is by some supposed to be a corruption of *folk's glove*, that is *Fairies' gloves*. The powdered leaf is a valuable medicine in cases where it is desired to lower the pulse.—Fl. June, July. Biennial.

ANTIRRHINUM ORONTIUM (*Lesser Snapdragon*).

2. ANTIRRHINUM (*Snapdragon*).

1. *A. majus* (Great Snapdragon).—*Leaves* narrow, tapering; *spikes* many-flowered; *sepals* egg-shaped,

blunt, much shorter than the corolla.—In limestone-quarries, chalk-pits, and on old walls, common. A handsome plant, with numerous leafy stems, each of which bears a spike of large, erect, personate flowers of a purple hue sporting to rose-colour or white. Specimens are common in gardens, the tints of which vary considerably; the most beautiful is of a rich crimson; one of a delicate lemon-colour is also frequent. Children derive much amusement from pinching the flowers between the finger and thumb, when the palate opens, as if in imitation of the fabulous monster from which it derives its name.—Fl. June—August. Perennial.

2. *A. Oróntium* (Lesser Snapdragon).—*Leaves* very narrow, tapering; *spikes* few-flowered; *sepals* much longer than the corolla.—Corn-fields, not uncommon. Smaller than the last, and at once distinguished by its leafy sepals, which are much longer than the small purple flowers.—Fl. July—September. Annual.

3. LINARIA (*Toad-flax*).

1. *L. vulgáris* (Yellow Toad-flax).— *Leaves* linear, tapering to a point, crowded; *flowers* in dense spikes; *sepals* smooth, shorter than the spur or capsule.—Hedges, very common. An erect herbaceous plant, 1—2 feet high, with numerous grass-like leaves of a glaucous hue, and dense spikes or clusters of yellow flowers, which are shaped like those of the *Snapdragon*, but spurred at the base. A variety is sometimes found with a regular corolla and five spurs.—Fl. August, September. Perennial.

2. *L. Elátiné* (Sharp-pointed Fluellen, or Toad-flax).—*Leaves* halbert-shaped; *stem* trailing.—Corn-fields, frequent. A small prostrate plant, with downy stems and leaves; solitary, axillary flowers, of which the upper lip is deep purple, the lower, yellow.—Fl. July—September. Annual.

LINARIA VULGARIS (*Yellow Toad-flax*), *and* LINARIA ELATINE (*Sharp-pointed Fluellen, or Toad-flax*).

3. *L. spuria* (Round-leaved Toad-flax).—*Leaves* egg-shaped ; *stem* trailing.—Corn-fields, not general. Resembling the last so closely, that it might be mistaken for a luxuriant specimen. The flowers are of the same colour, but larger ; the leaves, however, are always rounded at the base, not halbert-shaped.—Fl. July—September. Annual.

4. *L. Cymbalária* (Ivy-leaved Toad-flax).—*Leaves* heart-shaped, 5-lobed, smooth ; *stem* creeping.—On old garden walls, common. Not a native species, but quite naturalized ; growing freely from seed, and extending widely, by help of its long, rooting stems. The flowers are small, solitary, and light purple ; the leaves somewhat fleshy, and of a purple hue beneath. In the west of England it is commonly known by the name of *Mother-of-thousands.*—Fl. nearly all the year round. Perennial.

* Less common species of *Linaria* are *L. minor* (Least Toad-flax), a small, erect, much-branched plant, with very narrow, downy *leaves*, and purplish-yellow *flowers*: *L. repens* (Pale-blue Toad-flax), a slender, erect plant, 1—2 feet high, with glaucous, very narrow *leaves*, and veined, purplish-blue *flowers*; and *L. Itálica* (Italian Toad-flax), a very rare species with the habit of the last, but bearing yellow *flowers*.

SCROPHULARIA AQUATICA (*Water Fig-wort*).

4. SCROPHULARIA (*Fig-wort*).

1. *S. nodósa* (Knotted Fig-wort).—*Stem* square, with the angles blunt ; *leaves* smooth, heart-shaped, tapering to a point; *flowers* in loose panicles.—Moist bushy

places, common. A tall herbaceous plant, 3—4 feet
high, with repeatedly-forked panicles of almost globular,
dingy purple flowers, but attractive neither in form nor
colour.—Fl. June, July. Perennial.

2. *S. aquática* (Water Fig-wort).—*Stem* square, with
the angles winged ; *leaves* smooth, heart-shaped, oblong,
blunt ; *flowers* in close panicles.—Sides of streams and
ditches, common. Resembling the last, but at once distinguished by the winged angles of its stems, which,
though hollow and succulent, are rigid when dead, and
prove very troublesome to anglers, owing to their lines
becoming entangled in the withered capsules.—Fl. July,
August. Perennial.

* *S. Scorodonia* (Balm-leaved Fig-wort), is found
only in Cornwall, and at Tralee, in Ireland ; it is distinguished by its downy, wrinkled leaves : *S. vernális*
(Yellow Fig-wort), is a plant of local occurrence, and is
well distinguished by its remarkably bright-green foliage
and yellow flowers. It appears early in Spring, and is
the only British species which can be called ornamental.

LIMOSELLA AQUATICA (*Common Mud-wort*).

5. LIMOSELLA (*Mud-wort*).

1. *L. aquática* (Common Mud-wort).—The only
British species ; growing in watery places, but not

general.—A small plant, throwing up from the roots a number of smooth *leaves* on long stalks, and several minute, pale rose-coloured or white *flowers*, which are overtopped by the leaves.—Fl. July, August.   Annual.

MELAMPYRUM PRATENSE (*Common Yellow Cow-wheat*).

6. Melampýrum (*Cow-wheat*).

1. *M. pratensé* (Common Yellow Cow-wheat).—*Leaves*, in distant pairs, narrow, tapering, smooth ; *flowers*

axillary, in pairs, all turning one way; *corolla* four times as long as the calyx, *lower lip* longer than the *upper*.—Woods, common. A slender plant, about a foot high, with opposite pairs of straggling branches below, very narrow leaves, and long-tubed, yellow flowers. Cows are said to be fond of it, and, according to Linnæus, the best and yellowest butter is made where it abounds. The name *pratensé* (growing in meadows) was given to it erroneously, as it is never found in such situations.—Fl. June—August. Annual.

* *M. sylváticum* is a smaller species, occurring in mountainous woods, but not common. The *corolla* is only twice as long as the calyx, and the *lips* are equal.

2. *M. arvensé* (Purple Cow-wheat).—*Flowers* in oblong spikes; *bracts* leaf-like, very much cut and toothed.—Corn-fields in Norfolk and the Isle of Wight. Very distinct from the preceding, and well marked by its terminal spikes of yellow and purple flowers, which are almost buried among numerous bright rose-coloured bracts.—Fl. June, July. Annual.

* *M. cristátum* (Crested Cow-wheat), is distinguished from the preceding by bearing its flowers in 4-sided spikes. Both are beautiful plants.

7. PEDICULARIS (*Red-rattle*).

1. *P. palustris* (Marsh Red-rattle).—*Stem* solitary, erect, branched throughout; *calyx* downy, with 2 deeply-cut lobes. — Marshes and bogs, common. A herbaceous plant, well distinguished by the purple tinge of its branches, which are arranged in a pyramidal manner, its deeply-cut leaves, and large, crimson flowers. It is often a conspicuous plant in boggy ground, growing 12—18 inches high, and overtopping most of the surrounding herbage.—Fl. June—September. Perennial.

2. *P. sylvática* (Dwarf Red-rattle). — *Stems*, several from the same root, prostrate, unbranched; *calyx* smooth, with 5 unequal, leaf-like lobes.—Heathy places, common. Distinguished from the last by its humbler growth

and rose-coloured flowers, as well as by the above characters.—Fl. June—August. Perennial.

PEDICULARIS PALUSTRIS (*Marsh Red-rattle*).

8. RHINANTHUS (*Yellow-rattle*).

1. *R. Crista-galli* (Cock's-comb, Yellow-rattle). —
Leaves narrow-oblong, tapering to a point, serrated ;

RHINANTHUS CRISTA-GALLI (*Cock's-comb*, *Yellow-rattle*).

flowers in loose spikes; *bracts* egg-shaped, deeply serrated.
—In cultivated land, common. An erect, somewhat
rigid plant, 12—18 inches high, composed of a single
stem, and terminating in a loose spike of yellow flowers,
which are rendered conspicuous by their inflated calyces.
" When the fruit is ripe, the *seeds* rattle in the husky
capsule, and indicate to the Swedish peasantry the season
for gathering in their hay. In England, Mr. Curtis well
observes, hay-making begins when the plant is in full
flower." (Sir W. J. Hooker.)—Fl. June. Annual.

    * Another species or variety, *R. major* (Large bushy
Yellow-rattle), occurring in the north of England, bears
the *flowers* in crowded *spikes*, and at the base of each is
a yellowish bract, ending in a fine point.

9. BARTSIA.

1. *B. viscósa* (Yellow viscid Bartsia).—*Leaves* narrow, tapering, deeply serrated, *lower* opposite, *upper* alternate; *flowers* axillary.—Marshes and wet pastures, not common. Somewhat resembling *Rhinanthus Crista-galli* (Yellow-rattle), but at once distinguished by its solitary, not spiked, yellow flowers, and by being covered with clammy down.—Fl. August, September. Annual.

BARTSIA ODONTITES (*Red Bartsia*).

2. *B. Odontítes* (Red Bartsia).—*Leaves* narrow, tapering, serrated; *flowers* in one-sided, spike-like clusters.—Corn-fields, abundant. A much-branched herbaceous plant, 6—12 inches high, with narrow, dingy, purplish-green leaves, and numerous one-sided spikes of small

pink flowers. While flowering the spikes usually droop towards the ends.—Fl. July—September. Annual.

* *B. alpína* (Alpine Bartsia), is a rare species, found in Scotland and the north of England, and approaching *B. viscósa* in habit. In this species the *leaves* are all opposite, and the *flowers* grow in a short, leafy spike.

EUPHRASIA OFFICINALIS (*Common Eye-bright*).

10. EUPHRASIA (*Eye-bright*).

1. *E. officinalis* (Common Eye-bright).—The only British species. An elegant little plant, 2—6 inches high, with deeply-cut *leaves* and numerous white or purplish *flowers* variegated with yellow. On the mountains and near the sea, the stem is scarcely branched, and the leaves are fleshy ; but in rich soil it assumes the habit of a minute shrub. An infusion of this plant makes a useful eye-water.—Fl. July, August. Annual.

11. SIBTHORPIA (*Cornish Money-wort*).

1. *S. Europæa* (Cornish Money-wort).—The only species. An elegant little plant, clothing the banks of

SIBTHORPIA EUROPÆA (*Cornish Money-wort*).

springs and rivulets in most parts of Cornwall, and occasionally met with in some of the other southern counties. It approaches in habit *Hydrocótylé vulgaris* (Marsh Pennywort), but is smaller, and has downy, notched leaves. The *stems*, which creep along the ground in tangled masses, are slender and thread-like; the *leaves* small and of a delicate green; the *flowers* very minute and of a pale flesh-colour.—Fl. June—September.   Perennial.

12. VERÓNICA (*Speedwell*).

* *Flowers in terminal spikes or clusters.*

1. *V. serpyllifolia* (Thyme-leaved Speedwell). — *Leaves* egg-shaped, or elliptical, slightly notched, smooth; *capsules* inversely heart-shaped, with a long *style*.— Waste ground, common.   A small plant, with prostrate, or slightly ascending stems, and several many-flowered spikes of very light blue flowers, striped with dark blue veins.   A variety occurs, high up in the mountains,

which might easily be mistaken for a distinct species ; but on examination it will be found to differ only in the superior size of the flowers.—Fl. May—July. Perennial.

2. *V. arvensis* (Wall Speedwell).—*Leaves* egg-shaped, heart-shaped at the base, crenate, stalked ; *bracts* sessile, longer than the flowers.—Walls and fields, common. Also a small plant with inconspicuous, light blue flowers, which are almost concealed among the upper leaves or bracts. The whole plant is downy, and very frequently covered with dust.—Fl. April—September. Annual.

* To this group belong *V. spicata* (Spiked Speedwell), found on chalky pastures, about Newmarket and Bury ; *V. alpina* (Alpine Speedwell), a rare species, found only near the summits of the Highland mountains, distinguished from *V. serpyllifolia* by the deeper blue of its *flowers*, and by its very short *style*. *V. saxátilis*, (Blue Rock Speedwell), also a rare mountainous species, with slender, woody *stems*, large brilliant blue *flowers* and egg-shaped *capsules* : *V. fruticulosa* (Flesh-coloured Speedwell), found only on two of the Scotch mountains : *V. triphyllos*, also a rare species, growing in Norfolk and Suffolk, distinguished by its fingered *leaves*, and dark blue *flowers* : and *V. verna* (Vernal Speedwell), another rare species found only in Suffolk ; it approaches *V. arvensis* in habit, but has pinnatifid *leaves*.

* * *Clusters axillary.*

3. *V. Chamœdrys* (Germander Speedwell).—*Stem* with two hairy opposite lines ; *leaves* sessile, deeply serrated ; *clusters* very long ; *capsule* shorter than the calyx.— Hedge-banks, abundant. A well-known plant, which under the popular names of *Blue Speedwell* and *Bird's-eye*, is a favourite with every one. No one can have walked in the country in Spring without admiring its cheerful bright blue flowers, but few perhaps have remarked the singular pair of hairy lines, which traverse the whole length of the stem, shifting from side to side whenever they arrive at a fresh pair of leaves. It is

sometimes, but erroneously called, *Forget-me-not.*—Fl. May, June. Perennial.

VERONICA CHAMÆDRYS (*Germander Speedwell*), V. OFFICINALIS (*Common Speedwell*), V. SCUTELLATA (*Marsh Speedwell*), and V. BECCABUNGA (*Brooklime*).

4. *V. montána* (Mountain Speedwell).—*Stem* hairy all round ; *leaves* stalked ; *clusters* few-flowered ; *capsule* much longer than the calyx.—Woods, common. Approaching the last in habit, but well distinguished by the above characters, and by its smaller, light blue flowers.—Fl. May, June. Perennial.

5. *V. officinalis* (Common Speedwell).—Rough with short hairs ; *stem* creeping ; *leaves* elliptical, serrated ; *flowers* in spikes.—Heaths and dry pastures, common. A hairy plant with prostrate stems and erect spike-like clusters of blue flowers ; varying from 2 to 6 inches in length, according to soil. The leaves are astringent and

bitter, and are sometimes made into tea.—Fl. May—
August. Perennial.

6. *V. Beccabunga* (Brooklime).—Smooth ; *leaves* elliptical, blunt, slightly serrated ; *clusters* opposite ; *stem*
rooting at the base.—Brooks and ditches, common. A
succulent plant about a foot high, with rather large,
smooth leaves, and bright blue flowers, abounding in
situations favourable to the growth of Water-cresses and
Water-parsnep.—Fl. June—August. Perennial.

7. *V. Anagallis* (Water Speedwell).—Smooth ; *leaves*
narrow, tapering, serrated ; *clusters* opposite ; *stem* erect.—
Streams and ditches, common. Resembling the last, but
distinguished by its larger size, narrow leaves, erect
growth, and small flesh-coloured flowers. Fl. June—
August. Perennial.

8. *V. scutellata* (Marsh Speedwell).—Smooth ; *leaves*
linear, slightly toothed ; *clusters* alternate ; *fruitstalks*
bent back ; *capsules* of 2 flat round lobes.—Marshes,
not uncommon. A weak straggling plant, well distinguished by its very narrow leaves, and large flat capsules. Flowers pale pink.—Fl. June—August. Perennial.

* * * *Flowers solitary, axillary.*

9. *V. hederifolia* (Ivy-leaved Speedwell). — *Leaves*
stalked, 5—7 lobed; *sepals* heart-shaped, fringed ; *capsule* of two swollen lobes.—A common weed everywhere,
bearing in the axil of each leaf a pale blue flower, the
stalk of which is bent back when in fruit. The capsule is
composed of 2 much swollen lobes, each of which contains 2 large black seeds.—Fl. all the summer. Annual.

10. *V. agrestis* (Field Speedwell).—*Leaves* stalked,
heart-shaped, deeply serrated ; *sepals* oblong, blunt ;
flower-stalks as long as the leaves.—A common weed,
with several long prostrate stems and small blue flowers.
The capsule is composed of 2 swollen, keeled lobes, and
each cell contains about 6 seeds.—Fl. all the summer.
Annual.

* Closely allied to the preceding are *V. polita*, distinguished by its small *leaves*, which are shorter than the *flower-stalks :* and *V. Buxbaumii*, a stouter plant, with large bright blue *flowers*, and sharply keeled *capsules*.

13. VERBASCUM (*Mullein*).

1. *V. Thapsus* (Great Mullein).—*Leaves* woolly on both sides, running down the stem ; *stem* simple ; *flowers* in dense spikes.—Road-sides, common. A stout herbaceous plant 2—5 feet high, remarkable for its large flannel-like leaves and club-shaped spikes of yellow flowers. Two of the 5 stamens are longer than the rest, and hairy, the remaining three are smooth. This plant, together with Burdock and Fox-glove, is often introduced by painters into the foreground of landscapes. Fl. July, August. Biennial.

2. *V. nigrum* (Dark Mullein).—*Leaves* oblong, heart-shaped, stalked, downy on both sides, especially below ; *flowers* in dense tufts on a long crowded spike.—Hedges and road-sides, but of local occurrence. A handsome plant, not so stout as the preceding, and of a darker hue. The flowers, which are very numerous, are bright yellow, and the stamens are covered with purple hairs. Fl. July—September. Biennial.

3. *V. Blattária* (Moth Mullein).—*Leaves* oblong, embracing the stem, smooth ; *flowers* in loose tufts on a long interrupted spike.—Banks, rare, except in the West of England, where it is not unfrequent. A tall and somewhat slender plant, with shining, crenate leaves, the lowest of which are often lobed at the base, and with large, very handsome, yellow flowers. The stamens are covered with purple hairs.—Fl. July, August. Biennial.

* *V. virgátum* (Primrose-leaved Mullein), which is allied to the preceding, has the lower leaves downy ; rare : *V. Lychnítis* has small cream-coloured flowers, and is chiefly found on a chalky soil : and *V. floccosum* is remarkable for the mealy down which clothes both sides of its leaves; it is found principally in Norfolk and Suffolk.

VERBASCUM THAPSUS (*Great Mullein*).

Ord. LXII.—LABIATÆ.—The Labiate Tribe.

Calyx tubular, regular, or 2-lipped ; *corolla* irregular, mostly 2-lipped (*labiate*), the lower lip largest, and 3-lobed; *stamens* 4, 2 longer than the others, or sometimes wanting ; *ovary* deeply 4-lobed ; *style* 1 ; *stigma* 2-cleft ; *fruit* of 4 seeds, each of which is enclosed within a distinct shell, or rind.—A large and strongly-marked natural order, comprising upwards of 2,000 species, which agree in having square stems, opposite leaves, labiate or 2-lipped flowers, and a 4-lobed ovary with a single style arising from the base of the lobes. They are most abundant in temperate climates, and are remarkable for not possessing injurious properties in any single instance. Many are fragrant and aromatic : *Patchouli* is a favourite perfume, both in its natural state and when distilled. Lavender contains a fragrant volatile oil, which is valued both for its fragrance, and as a medicine for its stimulant properties. Several kinds of mint, as Peppermint and Pennyroyal, are much used in medicine. Spear-mint, Basil, Thyme, Marjoram, Savory, and Sage, are commonly used as pot-herbs, furnishing both agreeable and wholesome condiments. Horehound, Ground-Ivy, and Balm, are in rural districts popular remedies for chest complaints. Rosemary is remarkable for its undoubted power of encouraging the growth of the hair, and curing baldness, and is the active ingredient in most good pomatums ; an infusion of it prevents the hair from uncurling in damp weather ; and it is one of the plants used in the preparation of Hungary water, and Eau de Cologne. The admired flavour of Narbonne honey is ascribed to the bees feeding on the flowers of this plant, as that of the honey of Hymettus is indebted for its flavour to Wild Thyme. Several species of Sage (*Salvia*) are also cultivated for the beauty of their flowers.

* *Stamens* 2.

1. LÝCOPUS (Gipsy-wort).—*Calyx* 5-toothed ; *corolla* 4-cleft, nearly regular. (Name in Greek signifying a *Wolf's-foot*, from a fancied resemblance in the leaves).

2. SALVIA (Sage).—*Calyx* 2-lipped ; *corolla* gaping ; *filaments* forked. (Name from the Latin *salvo*, to heal, from the healing properties of the genus.)

** *Stamens* 4.

† *Corolla nearly regular, its tube scarcely longer than the calyx.*

3. MENTHA (Mint).—*Calyx* equal, 5-toothed ; *corolla* 4-cleft, with a very short tube. (Name, the Latin name of the plant.)

†† *Corolla 2-lipped, lips nearly equal in length.*

4. THYMUS (Thyme).—*Calyx* 2-lipped, 10—13 ribbed, the throat hairy ; *Corolla* with the upper lip notched, the lower 3-cleft ; *flowers* in heads, or whorls. (Name, the Latin name of the plant.)

5. ORÍGANUM (Marjoram).—*Calyx* 5-toothed, 10—13 ribbed, the throat hairy ; *flowers* in spikes, which are imbricated with bracts. (Name from the Greek *oros*, a mountain, and *ganos*, joy, from the favorite station of the family.)

††† *Corolla with the upper lip very short, or wanting.*

6. ÁJUGA (Bugle).—*Calyx* 5-cleft ; *corolla* with a long tube, the upper lip very short, lower 3-cleft. (Name said to be corrupted from the Latin *Abiga*, an allied plant.)

7. TEUCRIUM (Germander).—*Calyx* 5-cleft ; *corolla* with the upper lip deeply 2-cleft, lower 3-cleft. (Name from *Teucer*, who is said to have been the first to use it in medicine.)

†††† *Corolla 2-lipped, lips unequal ; calyx 5-toothed ; stamens longer than the tube of the corolla.*

8. BALLÓTA (Black Horehound). — *Calyx* funnel-shaped with 5 sharp equal teeth ; *corolla* with the *upper lip* erect, concave; *lower* 3-lobed, the middle lobe largest, heart-shaped. (Name in Greek signifying *rejected*, from the offensive smell of the plant.)

9. LEONÚRUS (Motherwort).—*Calyx* with 5 prickly teeth ; *corolla* with the *upper lip* nearly flat, very hairy above; *anthers* sprinkled with hard, shining dots. (Name in Greek signifying a *Lion's tail*, from some fancied resemblance in the plant.)

10. GALEÓBDOLON (Weasel-snout).—*Calyx* with 5 ribs, and as many nearly equal teeth ; *corolla* with the *upper lip* arched, entire ; *lower* in 3 nearly equal acute lobes. (Name in Greek denoting that the plant has the *smell* of a *weasel*.)

11. GALEOPSIS (Hemp-nettle).—*Calyx* bell-shaped, with 5 prickly teeth ; *corolla* with an inflated throat ; *upper lip* arched, *lower* 3-lobed, with 2 teeth on its upper side. (Name in Greek, denoting that the flower bears some *resemblance* to a *weasel*.)

12. LAMIUM (Dead-nettle).—*Calyx* bell-shaped, with 10 ribs, and 5 teeth ; *corolla* with an inflated tube ; *upper lip* arched, *lower* 2-cleft, with 1 or 2 teeth at the base on each side. (Name from the Greek *laimos*, a throât, from the shape of the flower.)

13. BETÓNICA (Betony).—*Calyx* egg-shaped, with 10 ribs, and 5 sharp teeth ; tube of the *corolla* longer than the calyx ; *upper lip* slightly arched, *lower* flat, of 3 unequal lobes. (" Name altered from *Bentonic*, in Celtic ; *ben* meaning. head, and *ton*, good, or tonic."— *Sir W. J. Hooker.*)

14. STACHYS (Woundwort). — *Calyx* tubular, bell-shaped, with 10 ribs, and 5 equal teeth ; tube of the *corolla* as long as the calyx ; *upper lip* arched, *lower* 3-lobed, the side lobes bent back before withering.

(Name in Greek, signifying a *bunch*, from the mode of flowering.)

15. NÉPETA (Cat-mint). — *Calyx* tubular, oblique, 5-toothed ; tube of the *corolla* longer than the calyx ; *upper lip* flat, notched, *lower* 3-lobed, the middle lobe concave, notched, the side ones bent back before withering. (Name of doubtful origin.)

16. GLECHÓMA (Ground-Ivy).—*Calyx* tubular, with 5 teeth, of which the upper are longest ; tube of the *corolla* longer than the calyx ; *upper lip* flat, 2-cleft, *lower* 3-lobed, with the middle lobe flat, notched. (Name from the Greek *glechon*, an allied plant.)

††††† *Corolla 2-lipped, lips unequal ; calyx 5—10 toothed ; stamens shorter than the tube of the corolla.*

17. MARRÚBIUM (White Horehound).—*Calyx* with 5 or 10 teeth, the throat hairy ; tube of the *corolla* longer than the calyx ; *upper lip* straight, very narrow, deeply 2-cleft, *lower* 3-lobed. (Name of doubtful origin.)

†††††† *Corolla 2-lipped, the lips unequal ; calyx 2-lipped.*

18. ÁCINOS (Basil-Thyme).—*Calyx* tubular, swollen underneath ; *upper lip* 3-cleft, *lower* 2-cleft, throat hairy ; *flowers* in whorls, of about 6 together. (Name, the Greek name of some allied plant.)

19. CALAMINTHA (Calamint).— Resembling Ácinos, except that the *flowers* do not grow in *whorls*. (Name in Greek, signifying *good mint*.)

20. CLINOPODIUM (Wild Basil).—Resembling Ácinos, except that the *flowers* grow in branched *whorls*, among numerous bristle-shaped *bracts*. (Name from the Greek.)

21. MELITTIS (Wild Balm).—*Calyx* bell-shaped, much wider than the tube of the corolla, variously lobed; *upper lip* of the *corolla* nearly flat, entire, *lower* with 3 rounded, nearly equal lobes. (Name from the Greek *melitta*, a bee, from the quantity of honey contained in the tube.)

22. PRUNELLA (Self-heal). — *Calyx* flattened, and closed when in fruit ; *filaments* 2-forked. (Name from

a German word for the *quinsy*, for which complaint it
was considered a specific.)

23. SCUTELLARIA (Skull-cap). — *Upper lip* of the
calyx bulged outward about the middle, and finally
closing down like a lid over the fruit; tube of the
corolla much larger than the calyx. (Name from the
Latin *scutella*, a little cup, which the calyx somewhat
resembles.

LYCOPUS EUROPÆUS (*Common Gipsy-wort*).

1. Lycopus (*Gipsy-wort*).

1. *L. Europæus* (Common Gipsy-wort).—The only
British species.—On the banks of rivers and ditches,
frequent. An aquatic plant, with erect, scarcely branched
stems, 2 feet high, deeply cut, pointed, opposite *leaves*,
and small, pale, flesh-coloured *flowers*, growing in crowded
whorls in the axils of the upper leaves.—Fl. July,
August. Perennial.

SALVIA VERBENACA (*Clary, or Wild Sage*).

2. Salvia (*Sage*).

1. *S. Verbenáca* (Clary, or Wild Sage).—*Leaves* oblong, blunt, heart-shaped at the base, wavy at the edge,

and crenate ; *corolla* scarcely longer than the calyx.—
Dry pastures, especially near the sea, or on a chalky
soil. An aromatic herbaceous plant, 1—2 feet high,
rendered conspicuous by its long spikes of purple-blue
flowers, the calyx of which is much larger than the
corolla. The leaves are few, and much wrinkled, and
at the base of each flower are 2 heart-shaped, fringed,
acute bracts.—Fl. June—August. Perennial.

* *S. pratensis* (Meadow Clary) which is not considered a native plant, occurs in Kent, and is distinguished
by its *corolla* being twice as long as the calyx.

3. MENTHA (*Mint*).

1. *M. sylvestris* (Horse Mint).—*Leaves* egg-shaped,
tapering to a point, serrated, downy, hoary beneath;
flowers in a thick cylindrical spike ; *bracts* awl-shaped;
calyx very hairy.—Damp waste ground, frequent. A
strong-scented plant, usually growing in masses, with
downy foliage, very white beneath, and rather slender
spikes of lilac flowers, which are often interrupted below.
—Fl. August, September. Perennial.

2. *M. rotundifolia* (Round-leaved Mint). — *Leaves*
sessile, broadly elliptical, blunt, much wrinkled, nearly
smooth above, shaggy beneath ; *flowers* in dense cylindrical spikes.—Waste ground, not common. The spikes
in this species are more slender than in the last, the
stem is somewhat woody, and the leaves are much
wrinkled and remarkably blunt ; the scent is strong and
aromatic, but scarcely agreeable.—Fl. August, September. Perennial.

3. *M. hirsúta* (Hairy Mint).—*Leaves* stalked, egg-
shaped, serrated, downy ; *flowers* at the summit of the
stem in dense whorls, the highest whorls forming a
head.—Banks of rivers and marshes abundant. The
commonest of the mints, 1—2 feet high, growing in
extensive masses in wet places, and well distinguished
by its downy foliage, and whorls of lilac flowers, which,
towards the summit of the stem, are crowded into heads ;

the scent is strong and unpleasant.—Fl. August, September. Perennial.

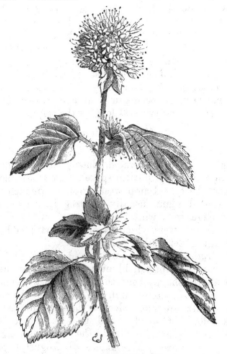

MENTHA HIRSUTA (*Hairy Mint*).

4. *M. arvensis* (Corn Mint).—*Leaves* stalked, egg-shaped, serrated, hairy ; *flowers* in dense, distant whorls ; *calyx* bell-shaped.—Corn-fields, common. A branched, downy plant, 6—12 inches high, with whorls of small lilac flowers, and a strong unpleasant smell.—Fl. Aug., September. Perennial.

5. *M. Pulégium* (Penny-royal). — *Stem* prostrate ; *leaves* egg-shaped, nearly smooth ; *flowers* in distant

whorls; *calyx* downy, its mouth closed with hairs.—
Wet heathy places, not common. The smallest of the
family, and very different in habit from any of the
others; the stems are prostrate, the flowers purple, and
the whole plant of an agreeable perfume and flavour.
It is commonly cultivated in cottage gardens for the
sake of being made into tea, which is a favourite remedy
for colds.—Fl. July, August. Perennial.

* Several other species and varieties of Mint are
described by botanists, some of which are scarcely distinct from the preceding; others, such as Pepper-Mint,
Spear-Mint, and Bergamot-Mint, are not really wild,
but have escaped from cultivation.

THYMUS SERPYLLUM (*Wild Thyme*).

4. Thymus (*Thyme*).

1. *T. Serpyllum* (Wild Thyme).—The only British
species.—Dry heathy places, common.   A well-known
and favourite little plant, with woody stems, small
fringed leaves, and heads of purple flowers.   The whole
plant diffuses a fragrant aromatic perfume, which, especially in hot weather, is perceptible at some distance.

ORIGANUM VULGARE (*Common Marjoram*).

5. Oríganum (*Marjoram*).

1. *O. vulgáré* (Common Marjoram).—The only British species.—Dry bushy places, especially on chalk or lime-stone, frequent. Growing about a foot high, and dis-tinguished by its egg-shaped downy *leaves*, and heads of purple *flowers*, which are crowded into the form of a *cyme*. The *bracts* are longer than the flowers, and tinged of the same colour, both being, while the plant is in bud, of a deep red hue. The whole plant is fra-grant and aromatic, and is frequently cultivated as a pot-herb.—Fl. July, August. Perennial.

AJUGA REPTANS (*Common Bugle*).

6. Ajuga (*Bugle*).

1. *A. reptans* (Common Bugle).—*Stem* erect, with creeping *scions* at the base; *lower leaves* stalked, *upper*

sessile ; *flowers* whorled, crowded into a spike.—Moist meadows and woods, common. Well marked by its solitary tapering flower-stalk, 6—9 inches high, and creeping scions. The flowers are blue, and the upper leaves, or bracts, are tinged with the same colour. A white variety is sometimes found.—Fl. May, June. Perennial.

2. *A. Chamæpitys* (Ground Pine). — *Stem* much branched, spreading ; *leaves* hairy, deeply 3-cleft, the segments linear ; *flowers* solitary, axillary. — Sandy fields, in Kent, Essex, and Surrey. A tufted herbaceous plant, 4—6 inches high, with reddish-purple viscid stems, finely cut leaves, and yellow flowers spotted with red. Its habit is very different from that of the preceding.— Fl. May, June. Perennial.

* *A. pyramidalis* (Pyramidal Bugle) is a rare Highland species, distinguished from common Bugle by being without *scions,* and by bearing its *whorls* of flowers crowded into 4-sided *spikes.*

7. TEUCRIUM (*Germander*).

1. *T. Scorodónia* (Wood-Germander, Wood-Sage).— *Stem* erect ; *leaves* heart-shaped, oblong, stalked, wrinkled ; *flowers* in 1-sided, spike-like clusters.—A common woodland plant, about 2 feet high, with Sage-like leaves, and several 1-sided clusters of small greenish-yellow flowers. The whole plant is very bitter, and has been used as a substitute for hops.—Fl. June—August. Perennial.

* *T. Scórdium* (Water Germander) is a rare species, growing in marshy places. It is only a few inches high, and bears its *flowers,* which are purple, in distant *whorls.* This plant was formerly employed in medicine as a tonic, and a protection against infectious diseases ; now, however, it is scarcely used except by rustic practitioners. *T. Chamædrys* is a doubtful native, and is also rare ;

the *flowers* are purple, with dark lines, large and handsome, and grow 3 together in the axils of the leaves. Several other species are frequently cultivated in gardens as ornamental plants.

TEUCRIUM SCORODONIA (*Wood Germander, Wood Sage*).

BALLOTA NIGRA (*Black Horehound*).

8. BALLÓTA (*Black Horehound*).

1. *B. nigra* (Black Horehound).—The only British species.—Waste ground, common.   A tall bushy plant, with downy, wrinkled, crenate *leaves*, and numerous purple *flowers*.   The odour of the whole plant is peculiarly strong and offensive.—Fl. July, September.   Perennial.

LEONURUS CARDIACA (*Common Motherwort*).

9. LEONÚRUS (*Motherwort*).

1. *L. Cardíaca* (Common Motherwort).—The only British species.—Hedges and waste places, not common. Distinguished from all other British plants of the Order by its *leaves*, which are deeply cut into 5 or 3 narrow, pointed segments, and by the prickly *calyx-teeth* of its *flowers*, which grow in whorls. When not in flower it resembles Mugwort (*Artemisia vulgaris*) in habit. The

stems are 2—3 feet high, branched principally below ; the upper *leaves* are very narrow and entire ; the *flowers* light purple.—Fl. August.　Perennial.

GALEOBDOLON LUTEUM (*Yellow Weasel-snout, Archangel, Yellow Dead-nettle*).

10. Galeóbdolon (*Weasel-snout.*)

1. *G. lúteum* (Yellow Weasel-snout, Archangel, Yellow Dead-nettle).—The only species.—Damp woods and hedges ; not unfrequent.　Resembling in habit the common White Dead-nettle, but rather taller ; the *leaves*

are narrow and more pointed ; the *flowers*, which grow in *whorls*, and are large and handsome, are yellow, blotched with red.—Fl. May—July. Perennial.

GALEOPSIS TETRAHIT (*Common Hemp-nettle*).

11. GALEOPSIS (*Hemp-nettle*).

1. *G. Tetráhit* (Common Hemp-nettle).—*Stem* bristly, swollen below the joints; *leaves* bristly, serrated.—Corn-fields, common. An erect, slender plant, 2 feet high, with opposite spreading branches, numerous whorls of flowers which are variegated with light purple and yellow,

and sometimes nearly white. The stems are remarkably swollen beneath every pair of leaves, and the whorls
of flowers are rendered conspicuous by the long sharp
calyx-teeth.—Fl. July—September. Annual.

2. *G. Ládanum* (Red Hemp-nettle).—*Stem* and *leaves*
downy with soft hair; *stem* not swollen below the
joints.—Gravelly and sandy fields; not unfrequent. Resembling the last, but smaller. The flowers are purple, mottled with crimson.—Fl. August—September.
Annual.

* *G. versicolor* (Large-flowered Hemp-nettle) is a plant
of local occurrence, resembling *G. Tetráhit* in character;
the *flowers* are large, yellow, with usually a broad purple
spot upon the lower lip : *G. villósa* (Downy Hemp-nettle)
is more like *G. Ládanum,* the *leaves* being soft and downy,
and the *stem* not swollen ; the *flowers* are large, pale
yellow.

12. Lámium (*Dead-nettle*).

1. *L. album* (White Dead-nettle).—*Leaves* heart-
shaped, tapering to a point, serrated, stalked.—Hedges
and waste ground, abundant. A common, but not inelegant weed, well marked by its large pure white flowers
and black stamens. So closely does the foliage of this
plant resemble that of the Stinging Nettle, that many
persons are afraid to handle it, supposing it to be a Nettle
in flower. The flowers of the latter, however, are green,
and so small, that they would be passed unnoticed but
for their growing in spiked panicles near the summit of
the stem. The square stem of the Dead-nettle is enough
to distinguish it at any stage of its growth.—Fl. all the
summer. Perennial.

2. *L. purpureum* (Purple Dead-nettle).—*Leaves* heart
or kidney-shaped, blunt, crenate, stalked.—A common
weed in cultivated ground, and by way-sides, distinguished
by the purple tinge of its foliage, crowded upper *leaves,*

and small purple *flowers*.—Fl. all the summer. Perennial.

LAMIUM PURPUREUM (*Purple Dead-nettle*).

* Allied to *L. album* is *L. maculatum* (Spotted Dead-nettle), distinguished by its *leaves*, each with a white blotch, and large purple *flowers* : two other species occur in similar situations with *L. purpureum*, and also resemble it in habit; namely, *L. amplexicaule* (Henbit-nettle), which has round, deeply-cut leaves, of which the upper are sessile: *L. incísum* (Cut-leaved Dead-nettle) the leaves of which are all deeply cut and stalked; both of these have small purple flowers.

13. BETÓNICA (*Wood Betony*).

1. *B. officinalis* (Wood Betony).—The only British species.—A common and very pretty woodland plant, about two feet high, bearing an interrupted *head* or *spike* of light purple *flowers* on a long and slender *stem*. There

are always 2 or 3 pairs of oblong crenate sessile *leaves* beneath the divisions of the spike ; the lower leaves are all stalked.—Fl. July, August.   Perennial.

BETONICA OFFICINALIS (*Wood Betony*).

14. STACHYS (*Wound-wort*).

1. *S. sylvática* (Hedge Wound-wort).—*Flowers* 6 in a whorl ; *stem* erect ; *leaves* heart-shaped, acute, stalked.— Woods and hedges, common.   A branched hairy plant, 2—3 feet high, with numerous whorls of dark but dull purple flowers, almost forming a spike.   There are rarely more or less than 6 flowers in a whorl, and the outline

of the leaves is not at all oblong. When in seed the
calyx-teeth are rigid.—Fl. July—August. Perennial.

STACHYS SYLVATICA (*Hedge Wound-wort*).

2. *S. palustris* (Marsh Wound-wort).—*Flowers* 6 in a whorl ; *leaves* narrow-oblong, heart-shaped at the base, sessile.—Marshes, common. Taller and stouter than the last, and distinctly marked by its oblong leaves tapering to a point, and light purple flowers.—Fl. July—August. Perennial.

3. *S. arvensis* (Corn Wound-wort).—*Flowers* 6 in a whorl ; *stem* spreading ; *leaves* heart-shaped, obtuse; *corolla* scarcely longer than the calyx.—Corn-fields, common. A small plant, 6—8 inches high, occurring abundantly as a weed in cultivated land ; distinguished from the preceding by its smaller size, and from the other *labiate* flowers which grow in such situations, by its whorls of 6 light purple flowers.—Fl. July—September. Annual.

* *S. ambigua* (Ambiguous Wound-wort) approaches *S. palustris*, from which it is distinguished by its *stalked leaves ;* it is of local occurrence, but is said to be abundant in the Highlands ; *S. Germánica*, (Downy Wound-wort), is a woolly plant having many flowered *whorls :* it is found on a chalky soil in Oxfordshire, Bedfordshire, and Berkshire.

15. Népeta (*Cat-mint*).

1. *N. Cataria* (Cat-mint).—The only British species. —Hedges and waste ground, not common. *Stem* branched, 2—3 feet high, white with mealy down, the *leaves* also are whitish beneath ; the *flowers*, which are small and whitish, dotted with crimson, grow in dense *whorls*, which towards the summit of the stem are so close as almost to form a *spike*. The whole plant has a strong aromatic odour, resembling Penny-royal, and peculiarly grateful to cats ; whence it derives its name. —Fl. July, August. Perennial.

NEPETA CATARIA (*Cat-mint*).

GLECHOMA HEDERACEA (*Ground Ivy*).

16. GLECHÓMA (*Ground Ivy*).

1. *G. hederacea* (Ground Ivy).—The only British species.—Hedges and waste ground, abundant. A favourite spring flower, with creeping *stems*, kidney-shaped, crenate, roughish *leaves*, and bright purple-blue *flowers* which grow in threes in the axils of the leaves. The whole plant has a strong aromatic odour, which, though scarcely fragrant, is far from disagreeable. In rural districts the leaves are often dried and made into tea.—Fl. April—June. Perennial.

17. MARRUBIUM (*White Horehound*).

1. *M. vulgáré* (White Horehound).—The only British species.—Waste ground not common. Well distinguished by its bushy *stems* 1—2 feet high, which are covered with white woolly down, by its wrinkled *leaves*,

and its dense whorls of small white *flowers*, of which the *calyx-teeth* are sharp and hooked. The whole plant is aromatic and bitter, and is a common remedy for coughs.—Fl. August. Perennial.

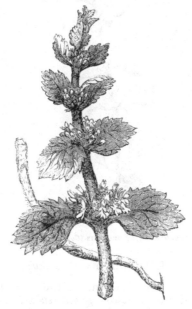

MARRUBIUM VULGARE (*White Horehound*).

18. ACINOS (*Basil-Thyme*).

1. *A. vulgáris* (Basil-Thyme).—The only British species.—Dry gravelly places, not common.—A small bushy herb 6—8 inches high, with hairy, egg-shaped *leaves* and purple *flowers*, which grow in *whorls* as well as at the summit of the *stem*. The *calyx* is distinctly 2-lipped, the *lower lip* bulged at the base.—Fl. July, August. Perennial.

CALAMINTHA OFFICINALIS (*Common Calamint*).

19. CALAMINTHA (*Calamint*).

1. *C. officinális* (Common Calamint).—*Leaves* egg-shaped, slightly serrated; *flowers* stalked, in forked axillary *cymes.*—Way-sides and hedges, not uncommon. An erect, bushy plant, with downy stems and foliage, and numerous, light purple flowers, which have small pointed bracts in the forks of their stalks. The whole plant has a sweet aromatic flavour and makes a pleasant tea.—Fl. July, August. Perennial.

* *C. Népeta* (Lesser Calamint), is a smaller plant, with *leaves* which are more strongly serrated, and bears its flowers on longer stalks. It is perhaps scarcely distinct from the last.

CLINOPODIUM VULGARE (*Wild Basil*).

20. CLINOPODIUM (*Wild Basil*).

1. *C. vulgare* (Wild Basil).—The only British species.
—Bushy places, frequent. A straggling hairy plant,
1—2 feet high, with egg-shaped *leaves*, several bristly
whorls of stalked, purple *flowers*, and numerous, long,
pointed *bracts*.—Aromatic and fragrant.—Fl. July,
August. Perennial.

21. MELITTIS (*Wild Balm*).

1. *M. Melissophyllum* (Wild Balm).—The only British
species.—Woods in the south and west of England.
A very handsome plant, about a foot high, with large,
hairy, serrated *leaves*, and conspicuous white *flowers*
blotched with bright rose-colour. The foliage while

fresh has an offensive smell, but in drying acquires the flavour of new hay or Woodruff.—Fl. June, July. Perennial.

MELITTIS MELISSOPHYLLUM (*Wild Balm*).

22. PRUNELLA (*Self-heal*).

1. *P. vulgaris* (Self-heal)—The only British species.— Pastures and waste ground, very common. Well distinguished by its flattened *calyx* and whorls of purplish-blue *flowers*, which are collected into a *head*, having a pair of *leaves* at the base, and two taper-pointed *bracts* beneath each *whorl*.—Fl. July, August. Perennial.

PRUNELLA VULGARIS (*Self-heal*).

23. SCUTELLARIA (*Skull-Cap*).

1. *S. galericuláta* (Greater Skull-Cap).—*Leaves* oblong, tapering, heart-shaped at the base, notched ; *flowers* in pairs, axillary.—Banks of rivers and ponds, frequent. A handsome plant 12—18 inches high, with rather large, bright blue flowers, the tube of which is much longer than the calyx. Soon after the corolla has fallen off the upper lip of the calyx closes on the lower, and gives it the appearance of a capsule with a lid; when the seed is ripe it opens again.—Fl. July—September. Perennial.

* *S. minor* (Lesser Skull-Cap), is a small bushy herb, 4—6 inches high, with egg-shaped *leaves*, of which the lower ones are often toothed at the base; the *flowers* are small, of a dull purple colour; the calyx is the same as

in the last. It grows in bogs, but is not common, except in the west of England.

SCUTELLARIA MINOR (*Lesser Skull-cap*).

Ord. LXIII.—VERBENACEÆ.—The Verbena Tribe.

Calyx tubular, not falling off; *corolla* irregular, with a long tube; stamens 4; 2 longer than the others, rarely 2 only; *ovary* 2- or 4-celled; style 1; *stigma* 2-cleft; *seeds* 2 or 4, adhering to one another.—A tribe of plants, closely allied to the *Labiatæ*, comprising trees, shrubs and herbaceous plants, having opposite leaves, and irregular flowers, which usually grow in spikes or heads. Many are aromatic and fragrant, and some few are employed as medicines, but are not highly valued. Great virtues were, in ancient times, attributed to the

VERBENA OFFICINALIS (*Common Vervain*).

common *Vervain*, insomuch that it was accounted a holy plant, and was used to sweep the tables and altars of the gods. It is now little thought off. *Aloysia citriodóra*, formerly called *Verbena triphylla*, is the Lemon-plant of gardens, well known for the delicious fragrance of its rough, narrow leaves. Many varieties of *Verbéna* are also cultivated for the sake of their ornamental flowers, which for brilliancy of colouring are scarcely surpassed. But by far the most remarkable plant in this Order is the Teak-tree (*Tectoria grandis*), which inhabits the mountainous parts of eastern Asia. The trunk of this tree sometimes attains the height of two hundred feet, and its leaves are twenty inches long by sixteen broad. The timber abounds in particles of flint, and somewhat resembles mahogany in colour, but is lighter and stronger. For ship-building it is thought to be superior to Oak.

1. VERBÉNA (Vervain).—*Calyx* 5-cleft; *corolla* un-equally 5-cleft; *stamens* shorter than the tube of the corolla. (Name, the Latin name of the plant.)

1. VERBÉNA (*Vervain*).

1. *V. officinalis* (Common Vervain).—The only British species.—Waste ground, common. A slender plant, 1—2 feet high, with but few *leaves*, which are roughish 3-cleft, or simply cut. The *flowers*, which are very small, are lilac, and grow in terminal very slender *spikes*.—This plant was held in great veneration by the ancients, being used in sacrifices and at other religious ceremonies. The cultivated species are showy plants, remarkable for the brilliant colours of their flowers. They are readily propagated by cuttings, which should be planted in fine sand during autumn, and protected during winter. New varieties are constantly being raised from seed.—Fl. July, August. Perennial.

Ord. LXIV.—LENTIBULARIÆ.—Butterwort Tribe.

Calyx divided, not falling off; *corolla* irregular, 2-lipped, spurred; *stamens* 2, sometimes 4, 2 long and 2 short; *ovary* 1-celled; *style* 1, very short; *stigma* 2-lipped, the lower lip smallest; *capsule* 1-celled, 2-valved, many seeded.—Herbaceous aquatic plants, bearing either undivided leaves, which spring directly from the root, or compound root-like leaves, with numerous small bladders or air-vessels. There are but four genera in the Order, two of which contain British examples; Butterwort *(Pinguicula)*, small plants with handsome purple flowers and concave leaves, of a texture which resembles greasy parchment; and Bladderwort *(Utricularia)*, submersed plants with finely divided leaves, bearing minute bladders, and yellow flowers, which rise above the surface of the water to open. "*Pinguicula vulgáris* (Common Butterwort), has the property of giving consistence to milk, and of preventing its separating into either whey or cream. Linnæus says that the solid milk of the Laplanders is prepared by pouring it, warm from the cow, over a strainer on which fresh leaves of Pinguicula have been laid. The milk, after passing among them, is left for a day or two to stand, until it begins to turn sour; it throws up no cream, but becomes compact and tenacious, and most delicious in taste. It is not necessary that fresh leaves should be used after the milk is once turned: on the contrary, a small portion of this solid milk will act upon that which is fresh, in the manner of yeast." (*Lindley*).

1. Pingufcula (Butterwort).—*Calyx* 2-lipped, *upper lip* 3-cleft, lower 2-cleft; *corolla* gaping, spurred. (Name from the Latin *pinguis*, fat, the leaves being greasy to the touch.)

2. Utricularia (Bladderwort).—*Calyx* of 2 equal sepals; *corolla* personate, spurred. (Name from the Latin *Utriculus*, a little bladder, from the little air-bladders which grow among the leaves.)

1. PINGUICULA (*Butterwort*).

1. *P. vulgáris* (Common Butterwort).—*Spur* tapering; segments of the *corolla* very unequal, entire.—Bogs and heaths, principally in the North. A singular and very beautiful plant. The leaves, which spring all from the roots, have the edges rolled in ; they are of a peculiar, parchment-like hue, and have a frosted appearance. The flowers are large, purple, very handsome, and grow in a nodding manner, each on the summit of a delicate stem, 3—4 inches high, which springs directly from the root. The root is fibrous, and has a very loose hold on the soft ground in which it grows.—Fl. June. Perennial.

PINGUICULA LUSITANICA (*Pale Butterwort*).

2. *P. Lusitánica* (Pale Butterwort),—*Spur* cylindrical, obtuse, curved downwards; segments of the *corolla* nearly equal; *leaves* and *flower-stalks* covered with short hairs.—Bogs in the western parts of England, in the North of Scotland, and in Ireland. Of the same habit as the last, but much smaller. The leaves are greenish white, and veined; the flowers light pink.—Fl. July—September. Perennial.

* *P. grandiflóra* (Large-flowered Butterwort), is a yet more beautiful plant than *P. vulgáris*. It is distinguished by its large very irregular purple *corolla*, the middle segment and spur of which are notched; it grows in bogs in the counties of Cork and Kerry, Ireland. *P. alpína* (Alpine Butterwort), is of about the size of *P. Lusitánica;* the *flower-stalks* are smooth, and the *flowers* yellowish; it is very rare, being found only in bogs in Scotland and Ireland.

2. UTRICULARIA (*Bladderwort*).

1. *U. vulgáris* (Common Bladderwort).—Submersed. *Leaves* divided into numerous hair-like segments, and bearing small air-bladders; *lips* of the *corolla* about equal in length; spur conical.—Ditches and deep pools, not very common. Before flowering, the stem and leaves float in the water by help of the minute bladders, which are then filled with air; the flowers, which grow in clusters of 6—8 together, are large, and bright yellow, and are raised several inches out of the water. After flowering the bladders become filled with water, and the whole plant sinks to the bottom.—Fl. June, July. Perennial.

* *U. minor* (Lesser Bladderwort), is a rare species, with small yellow *flowers*, and a short blunt *spur;* it is most frequent in Scotland: *U. intermedia* (Intermediate Bladderwort), is also a rare species, distinguished from the common one by having the *upper lip* of the corolla

much longer than the *lower*, and by bearing its air-
bladders on branched stalks distinct from the leaves.

UTRICULARIA VULGARIS (*Common Bladderwort*).

Ord. LXV.—PRIMULACEÆ.—Primrose Tribe.

Calyx 5-cleft, rarely 4-cleft, (in *Trientális* 7-cleft,) regular, not falling off ; *corolla* of as many lobes as the calyx (in *Glaux* wanting); *stamens* equalling in number the lobes of the corolla, and opposite to them ; *ovary* 1-celled ; *style* 1 ; *stigma* capitate; *capsule* 1-celled, opening with valves ; *seeds* numerous, attached to a central column.—Herbaceous plants, mostly of humble growth, inhabiting, principally, the colder regions of the northern hemisphere, and in lower latitudes ascending to the confines of perpetual snow. In this order are found several of our most favourite British plants. The Primrose, as its name indicates, *(prima rosa,* the first rose,) is the most welcome harbinger of spring ; the Cowslip is scarcely less prized for its pastoral associations than for its elegance and fragrance ; Pimpernel, or " Poor man's weather-glass," is as trusty a herald of summer weather as the Primrose of spring. Nor is it only as *Flowers of the Field* that the plants of this tribe are valued. The Polyanthus and Aurícula equally grace the cottager's garden, and the collections of the florist ; and several species of Cýclamen are commonly found in conservatories. Some species possess active medicinal properties ; the flowers of the Cowslip are made into a pleasant soporific wine ; the leaves of the Aurícula (*Prímula Aurícula*) are used in the Alps as a remedy for coughs ; and the flowers of Pimpernel and roots of Cýclamen are acrid.

1. Prímula (Primrose).—*Calyx* tubular, 5-cleft; *corolla* salver or funnel-shaped, with a long cylindrical tube; *stamens* 5, enclosed within the tube of the corolla ; *capsule* 5-valved with ten teeth. (Name from the Latin *primus,* first, from the early appearance of the flowers.)

2. Hottonia (Water Violet).—*Calyx* 5-cleft almost to the base ; *corolla* salver-shaped, with a short tube ; *stamens* 5 ; *capsule* opening with 5 teeth. (Named after Professor Hotton, of Leyden.)

3. Cÿclamen (Sow-bread).—*Calyx* bell-shaped, cleft half way down into 5 segments ; *corolla* wheel-shaped, the lobes reflexed ; *stamens* 5 ; *capsule* opening with 5 teeth. (Name from the Greek *cyclos,* a circle, either from the reflexed lobes of the corolla, or from the spiral form of the fruit-stalks.)

4. Anagallis (Pimpernel).—*Calyx* 5-cleft to the base ; *corolla* wheel-shaped ; *stamens* 5, hairy ; *capsule* splitting all round. (Name in Greek denoting that the plant excites pleasure.)

5. Lysimachia (Loosestrife).—*Calyx* · 5-cleft to the base ; *corolla* wheel-shaped ; *stamens* 5, not hairy ; *capsule* opening by valves. (Name in Greek having the same meaning as the English name.)

6. Centúnculus (Chaffweed).—*Calyx* 5-cleft to the base ; *corolla* with an inflated tube ; *stamens* 4 ; *capsule* splitting all round. (Name of doubtful etymology.)

7. Trientális (Chickweed Winter-green).—*Calyx* 7-cleft to the base ; *corolla* wheel-shaped ; *stamens* 7 ; *capsule* opening with valves. (Name of doubtful etymology.)

8. Glaux (Sea-milkwort).—*Calyx* 0 ; *corolla* bell-shaped, 5-lobed ; *stamens* 5 ; *capsule* 5-valved, with 5— 10 seeds. (Name in Greek denoting the sea-green colour of the foliage.)

9. Sámolus (Brookweed).—*Calyx* 5-cleft, adhering to the lower half of the capsule, not falling off ; *corolla* salver-shaped, with 5 scales at the mouth of the tube ; *stamens* 5 ; *capsule* opening with 5 reflexed teeth. (" Named, some say, from the Island of Samos, where Valerandus, a botanist of the 16th century, gathered our *Samolus Valerandi.*")—*Sir W. J. Hooker.*

1. Prímula (*Primrose*).

1. *P. vulgáris* (Primrose).—*Flowers* each on a separate stalk ; *leaves* oblong, egg-shaped.—Banks and woods, abundant. Among the most welcome of spring flowers, and too well known to need any description. The

colour of the flower is so peculiar as to have a name
of its own ; artists maintain that primrose-colour is a
delicate green ; white, purple, and lilac varieties are not
uncommon.—Fl. March—May.   Perennial.

PRIMULA VULGARIS (*Primrose*).

2. *P. elatior* (Oxlip).—*Flowers* in a stalked umbel, salver-shaped ; *calyx* tubular ; *leaves* egg-shaped, contracted below the middle.—Woods and pastures, not common. Distinguished from the *Primrose* by its umbellate yellow flowers, and by its leaves, which become suddenly broader above the middle, and from the Cowslip by its tubular, not bell-shaped calyx, and flat, not concave corolla.—Fl. April, May. Perennial.

3. *P. veris* (Cowslip, Paigle).—*Flowers* in a stalked umbel, drooping, funnel-shaped; *calyx* bell-shaped; *leaves* egg-shaped, contracted below the middle.— Pastures, common. Among the many pleasing purposes to which these favourite flowers are applied by children, none is prettier than that of making *Cowslip Balls*. The method, which may not be known to all my readers, is as follows: —The umbels are picked off as close as possible to the top of the main stalk, and from fifty to sixty are made to hang across a string stretched between the backs of two chairs. The flowers are then carefully pressed together, and the string is tied tightly so as to collect them into a ball. Care should be taken to choose such heads or umbels only as have all the flowers open, or the surface of the ball will be uneven.—Fl. April, May. Perennial.

* *P. farinósa* (Bird's-eye Primrose), is found in mountainous pastures in the North of England, and South of Scotland ; its leaves are thickly covered with a powdery meal, and the flowers are purple with a yellow eye : *P. Scótica* (Scottish Primrose), is a rare species, growing in the Orkneys, and in a few places in the North of Scotland ; it resembles the last, but is smaller. All the varieties of Primrose, Cowslip, and Auricula, may be easily propagated by dividing the roots in autumn. New varieties are raised from seed, which should be sown as soon as ripe in leaf-mould, and pricked out into beds when large enough.

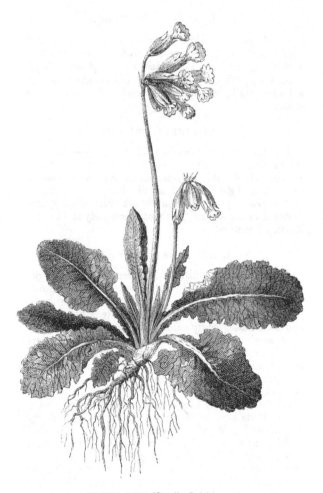

PRIMULA VERIS (*Cowslip. Paigle*).

2. HOTTONIA (*Water-Violet*).

1. *H. palustris* (Water-Violet).—The only British species.—Ponds and ditches, frequent. An aquatic plant, with finely divided, submersed *leaves ; flowers* large, handsome, pink and yellow, arranged in *whorls* around a leafless stalk, which rises several inches out of the water.

3. CYCLAMEN (*Sow-bread*).

1. *C. hederæfolium* (Ivy-leaved Sow-bread).—The only species found in Britain, and probably not a native.— Remarkable for its globular, brown *root*, and nodding flesh-coloured *flowers*, the lobes of which are bent back. As the *fruit* ripens, the flower-stalk curls spirally, and buries it in the earth. The root is intensely acrid.—Fl. April. Perennial.

4. ANAGALLIS (*Pimpernel*).

1. *A. arvensis* (Scarlet Pimpernel).—*Leaves* egg-shaped, dotted beneath, sessile ; *petals* crenate.—Cultivated ground, abundant. A pretty little plant, with bright scarlet flowers, which expand only in fine weather and have consequently gained for the plant the name of *Poor man's weather-glass.* The colour of the flowers occasionally varies to flesh-coloured, or white with a red eye. A bright blue variety, which some botanists consider a distinct species, is more unfrequent.—Fl. June, July, Annual.

2. *A. tenella* (Bog Pimpernel).—*Stem* creeping ; *leaves* roundish, stalked, shorter than the flower-stalks.—Boggy ground and sides of rivulets, common. A beautiful little plant with slender stems 4—6 inches long, small leaves which are arranged in opposite pairs, and erect rose-coloured flowers, larger than those of the *Scarlet Pimpernel,* and more frequently having the lobes of the corolla erect than expanded.

ANAGALLIS ARVENSIS (*Scarlet Pimpernel*), and A. TENELLA (*Bog Pimpernel*).

5. LYSIMACHIA (*Loosestrife*).

1. *L. nummularia* (Money-wort, Herb-twopence).— *Stem* creeping; *leaves* roundish, slightly stalked; *flowers* solitary, axillary.—Banks of rivers and damp woods, common. A very pretty plant, well marked by its opposite, shining leaves, and large, yellow flowers. The stems often grow a foot or more in length, and hang from the banks of rivers in a very graceful way. This plant is much used to ornament artificial rockeries.—Fl. June, July. Perennial.

2. *L. némorum* (Wood Loosestrife, Yellow Pimpernel). — *Stem* spreading; *leaves* egg-shaped, acute; *flowers* solitary, axillary.—Woods, common. Approaching the Scarlet Pimpernel in habit, but somewhat larger; the flowers are bright yellow, and very pretty.—Fl. June, July. Perennial.

LYSIMACHIA NEMORUM (*Wood Loosestrife, Yellow Pimpernel*).

3. *L. vulgáris* (Great Yellow Loosestrife).—*Stem* erect, branched; *leaves* tapering to a point, opposite, or 3—4 in a whorl; *flowers* in terminal panicles.—Banks of rivers, common. Very different in habit from either of the preceding; growing quite erect, 2—3 feet high, with terminal panicles of rather large, yellow flowers.—Fl. July. Perennial.

* *L. thyrsiflóra* (Tufted Loosestrife), resembles *L. vulgáris* in habit, but bears its flowers, which are small and yellow, in numerous, dense clusters; it grows in marshes in Scotland and the North of England, but is rare.

6. CENTÚNCULUS (*Chaffweed*).

1. *C. mínimus* (Chaffweed).—The only British species. —One of the smallest among British plants, rarely exceeding an inch in height, and often much less. It is nearly allied to the *Pimpernel*, and at the first glance might be mistaken for a stunted specimen of the common species. The *leaves* are egg-shaped, acute; the *flowers* sessile, axillary. It is sometimes branched, but very frequently consists of a single stem, 6 or 8 leaves, and

as many inconspicuous flowers.—It grows in damp, gravelly places, especially where water has stood during winter.—Fl. June—August. Annual.

CENTUNCULUS MINIMUS (*Chaffweed*).

7. Trientális (*Chickweed Winter-green*).

1. *T. Europæa* (European Chickweed Winter-green). —The only British species, and the only British plant belonging to the Linnæan Class, Heptandria.—Abundant in many parts of the Highlands of Scotland, and occasionally found in the North of England. A pretty plant, with an unbranched *stem* 4—6 inches high, bearing a few large *leaves* near its summit, and one or more delicate white *flowers,* each on a slender stalk. The number of *stamens* varies from 7—9.—Fl. June. Perennial.

8. Glaux (*Sea-Milkwort*).

1. *G. marítima* (Sea-Milkwort).—The only species.— Sea shore and salt marshes, common. A fleshy, marine plant 3—6 inches high, growing in thick patches, with numerous egg-shaped, glaucous *leaves,* and axillary pink *flowers,* which are destitute of *calyx.* In habit it resembles *Arenaria peploídes.*—Fl. June—August. Perennial.

GLAUX MARITIMA (*Sea-Milkwort*).

9. SÁMOLUS (*Brookweed*).

1. *S. Valerandi* (Brookweed).— The only British species.—Watery places, common ; and like many other aquatic plants, widely diffused, being found in Africa, America, and New South Wales. A smooth, pale green, herbaceous plant, with blunt, fleshy *leaves*, and one or more terminal clusters of very small white *flowers*, which, in their early stage, are crowded, but finally become

distant, resembling in this respect the habit of the Cruciform Tribe.—Fl. July—September. Annual.

SAMOLUS VALERANDI (*Brookweed*).

Ord. LXVI—PLUMBAGINEÆ.—Thrift Tribe.

Calyx tubular, plaited, chaffy, not falling off, often coloured ; *corolla* 5-cleft nearly to the base ; *stamens* 5 opposite the petals, *ovary* of 5 carpels, 1-celled ; *styles* 5 ; fruit 1-seeded.—Herbaceous or somewhat shrubby plants, with undivided, fleshy leaves, and flowers of a thin texture, approaching that usually called *Everlasting*, collected into heads or growing in panicles.   They inhabit salt marshes and the sea shore of most temperate regions, and some are also found in mountainous districts. Their properties are various,—some are tonic, some intensely acrid, and many contain iodine.   The root of *Státicé Caroliniana* is one of the most powerful astringents known ; several species of *Plumbágo* are so acrid that the fresh root is used to raise blisters.   Thrift and several kinds of Sea-Lavender *(Státicé)* grow on the sea shores of Britain, and are very pretty plants.   Other species are cultivated in gardens and conservatories, to which they are highly ornamental.   It has been remarked that the plants of this order, like many other marine plants, when growing at a distance from the sea, lose the peculiar salts which they contain in their natural localities. Thrift, for example, as a marine plant contains iodine and soda, but as a mountain or garden plant, exchanges these two salts for potash.   The only British genus belonging to this tribe is

1. Státicé (Thrift).   Characters given above.   (Name from the Greek *statizo*, to stop, from its supposed medicinal virtues.)

1. Státicé (*Thrift*).

1. *S. Armeria* (Thrift).—*Leaves* linear, fleshy ; *flower-stalks* springing directly from the roots, leafless, and bearing each a roundish head of flowers.—Sea shores and the tops of mountains, common.   *Leaves* forming dense tufts

or balls ; *flowers* rose-coloured, in roundish heads, which are raised on downy stalks 3—6 inches from the ground ; the summit of the flower-stalk is cased in a brown membranous sheath, and the flowers are intermixed with chaffy bracts, or scales ; the fruit is almost winged by the dry, chaffy calyx.—Fl. July, August. Perennial.

STATICE SPATHULATA (*Spathulate Sea-Lavender*) and S. ARMERIA (*Thrift*).

2. *S. Limonium* (Sea-Lavender).—*Leaves* oblong 1-ribbed, tipped with a point ; *flower-stalk* from the root, leafless, branched near the summit into many spreading tufts.—Muddy sea coast, not unfrequent. Very different

from the last, having broad leaves and angular flower-stalks, which are branched at the summit into several spike-like clusters of thin, scentless flowers.—Fl. July, August. Perennial.

* *S. spathuláta* (Spathulate Sea-Lavender), is not uncommon on the rocky sea coast; it is distinguished by its *leaves* being oblong near the base and wider above (*spathulate*), and by its *flower-stalks* being branched below the middle into several erect tufts of blue flowers : *S. reticulata* (Matted Sea-Lavender), occurs only in the salt marshes of Norfolk ; the flower-stalks are divided almost from the base into numerous zigzag branches, of which the lower are barren.

ORD. LXVII—PLANTAGINEÆ.—PLANTAIN TRIBE.

Calyx 4-parted ; *corolla* 4-parted, chaffy, not falling off ; *stamens* 4, alternate with the segments of the corolla, and having very long, thread-like filaments, and lightly-attached anthers ; *ovary* 2—rarely 4-celled ; *style* 1 ; *stigma* hairy ; *capsule* splitting transversely ; *seeds* 1, 2, or many in each cell.—Herbaceous plants of humble growth, with many ribbed or fleshy leaves spreading horizontally from the root. The flowers, which are made conspicuous by their long stamens, grow in spikes. Several species are common in Great Britain as wayside, meadow, and marine plants, and many are found in almost all parts of the world. The leaves are slightly bitter and astringent ; the seeds abound in a tasteless mucilage, which is used in medicine as a substitute for Linseed, and is said to be employed in France to stiffen muslin.

1. PLANTÁGO (Plantain).—*Calyx* 4-cleft, the segments bent back ; *corolla* tubular, with 4 spreading lobes ; *stamens* very long ; *capsule* splitting all round, 2—4-celled. (Name of doubtful origin.)

2. LITTORELLA (Shore-weed).—Stamens and pistils in

different flowers: *barren flower* stalked ; *stamens* very long ; *fertile flower* sessile ; *bracts* 3, *corolla* tubular, contracted at both ends ; *style* very long ; *capsule* 1-seeded. (Name in Latin having the same meaning as the English name.)

1. PLANTAGO (*Plantain*).

1. *P. major* (Greater Plantain).—*Leaves* broadly egg-shaped on long channelled stalks ; *flowers* in a long spike, the stem of which is cylindrical ; *cells* of the *capsule* many-seeded.—Borders of fields and waysides, abundant. Well known for its spikes of green flowers, the seeds of which are a favourite food of Canary birds.—Fl. June, July. Perennial.

2. *P. media* (Hoary Plantain).—*Leaves* broadly ellip-tical on short flat stalks ; *flowers* in a close cylindrical spike, the stalk of which is also cylindrical ; *cells* of the *capsule* 1-seeded.—Meadows, common. The leaves spread horizontally from the crown of the root, and lie so close to the ground as to destroy all vegetation beneath, and to leave the impression of their ribs on the ground ; the spike is shorter than in *P. major*, but grows on a longer stalk, and the flowers, which are fragrant, are rendered conspicuous by their light purple anthers.—Fl. June, July. Perennial.

3. *P. lanceolata* (Ribwort Plantain).—*Leaves* narrow, tapering ; *flowers* in a short spike, the stalk of which is angular ; *cells* of the *capsule* 1-seeded.—Meadows, abun-dant. Under the name of *Cocks and Hens* this plant is well known to children, who amuse themselves by striking the heads one against another until the stalk breaks. The flowers are dark brown.—Fl. June, July. Peren-nial.

4. *P. maritima* (Sea-side Plantain).—*Leaves* linear, grooved, fleshy, woolly at the base.—Sea shores and tops of mountains, common. Easily distinguished from the

rest of the genus by its long, fleshy leaves.—Fl. June—
September.   Perennial.

5. *P. Corónopus* (Buck's-horn Plantain).— *Leaves*
pinnatifid ; *capsule* imperfectly 4-celled, 4-seeded.—
Waste ground, especially near the sea, common.—The
only British species which has divided leaves ; these are
more or less downy, and usually prostrate.—Fl. June
July.   Annual.

PLANTAGO LANCEOLATA (*Ribwort Plantain*).

2. LITTORELLA (*Shore-weed*).

1. *L. lacustris* (Shore-weed).—The only species.—
Marshes and banks of lakes. Not unlike *Plantago
maritima* in habit, but at once distinguished by its solitary *barren flowers*, raised each on a stalk 2—4 inches
high ; the *fertile flowers* are sessile among the leaves.—
Fl. June—September. Perennial.

LITTORELLA LACUSTRIS (*Shore-weed*).

SUB-CLASS IV.

THALAMIFLORÆ.

Flowers having a calyx or corolla, or neither, never both. In this Sub-class it is often doubtful whether the leaves which enclose the stamens and pistils of a flower should be called a *calyx* or *corolla :* the term *perianth* (from the Greek *peri,* around, and *anthos,* a flower) is therefore used to denote this organ, and must be taken to mean all the leaves, whether resembling *sepals* or *petals,* which enclose the other parts of fructification. Used in this sense, and applied to the preceeding Sub-classes, the calyx and corolla would be correctly called a *double perianth.*

ORDER LXVIII.—CHENOPODEÆ.—GOOSE-FOOT TRIBE.

Perianth 5-lobed, not falling off ; *stamens* 5, rarely 1 or 2, from the base of the perianth and opposite its lobes; *ovary* 1, superior or adhering to the tube of the perianth ; *style* 2 or 4-cleft, rarely simple ; *stigma* undivided ; *fruit* 1-seeded, enclosed in the perianth, which often becomes enlarged or fleshy —Herbaceous or somewhat shrubby plants, with leaves which are more or less inclined to be fleshy ; the flowers are small and inconspicuous, the perianth decidedly partaking of the characters of a *calyx,* which sometimes, as in *Atriplex,* has a tendency to become enlarged when in fruit. Some species have flowers bearing pistils only, others stamens only, and others again both stamens and pistils. They are common weeds in most temperate climates, and are most abundant in salt marshes and on the sea shore. Many of the plants of this tribe are used as esculent vegetables, as Spinach, Beet, and Orache. Beet is cultivated extensively in France for making sugar, and a variety of it affords valuable food for cattle under the name of Mangold Wurzel. In Peru the leaves of *Chenopodium Quinoa*, a plant growing at a great elevation, is a common article of food. Many of those kinds which grow in salt marshes and on the sea shore afford an immense quantity of soda. According to some naturalists *Salvadóra Persica*, belonging to this order, is the Mustard Tree of Scripture. It bears a juicy fruit, having the flavour of cress, and its seeds are very small. (For a further account of many of the plants belonging to this order, see Miss Pratt's "Common Things of the Sea-side.")

1. CHENOPÓDIUM (Goose-foot). — *Perianth* deeply 5-cleft, remaining unaltered, and finally closing over the single seed ; *stamens* 5 ; *stigmas* 2. (Name in Greek having the same meaning as the English name.)

2. ÁTRIPLEX (Orache).—*Stamens* and *pistils* for the most part in separate flowers, sometimes united ; *barren flower, perianth* deeply 5-cleft ; *stamens* 5 ; *fertile flower, perianth* of 2 valves ; *stigmas* 2 ; *fruit* 1-seeded, covered by the enlarged perianth. (Name from the Greek *a*, not, and *trephein*, to nourish.)

3. BETA (Beet).—*Perianth* deeply 5-cleft ; *stamens* 5 ; *stigmas* 2 ; *fruit* 1-seeded, adhering to the tube of the fleshy perianth. (Name, the Latin name of the plant.)

4. SÁLSOLA (Saltwort).— *Perianth* deeply 5-cleft ; *stamens* 5 ; *stigmas* 2 ; *fruit* 1-seeded, crowned by the shrivelled lobes of the perianth. (Name from the Latin *sal*, salt, from the alkaline salt in which it abounds.)

5. SALICORNIA (Glass-wort).—*Perianth* top-shaped, fleshy, undivided ; *stamens* 1—2 ; *style* very short ; *stigma* 2-cleft ; *fruit* enclosed in the dry perianth. (Name from the Latin *sal*, salt, and *cornu*, a horn, from the alkaline salt in which it abounds, and the horn-shaped branches.)

1. CHENOPODIUM *(Goose-foot)*.

* *Leaves semicylindrical ; flowers each with 2 bracts.*

1. *C. marítimum* (Annual Sea-side Goose-grass).— *Styles* 2 ; *stem* herbaceous.—Muddy sea-shore, common.

A low, straggling plant, with short, fleshy, semicylindrical leaves; and small, inconspicuous, green flowers.— Fl. July, August. Annual.

* To this group belongs also *C. fruticosum* (Shrubby Sea-side Goose-grass), a much rarer plant than the last, growing 2—3 feet high, with a shrubby *stem*, and having 3 *styles* in each flower.

* * *Leaves plane, undivided ; flowers without bracts.*

2. *C. ólidum* (Stinking Goose-foot).—*Stem* spreading ; *leaves* egg-shaped, with a triangular base, fleshy, mealy; *flowers* in dense clustered spikes.—Waste places, especially near the sea. Distinguished by its fishy smell, which is disgusting in the extreme.—Fl. August. Annual.

3. *C. polyspermum* (Many-seeded Goose-grass).—*Stem* spreading ; *leaves* egg-shaped, sessile ; *flowers* in branched, somewhat leafy, slender spikes ; *seeds* flattened horizontally, shining, minutely dotted.—Waste ground, not common. Varying in size, from 4 inches to a foot in height ; the stems and leaves usually have a red tinge, and the plant, when in flower, has a not inelegant appearance from the number of shining, brown seeds, which are not concealed by the perianth.

* * * *Leaves plane, toothed, angled, or lobed.*

4. *C. Bonus-Henricus* (Good King Henry).—*Leaves* triangular, arrow-shaped ; *flowers* in compound, leafless spikes.—Waste places near villages, common. A dark-green, succulent plant, about a foot high, with large, thickish leaves, which are used as *Spinach.*—Fl. August. Perennial.

5. *C. album* (White Goose-foot).—*Leaves* egg-shaped, with a triangular base, bluntly toothed, upper ones narrow, entire; *flowers* in dense, clustering spikes.—Waste places and cultivated ground, common. Whole plant succulent ; leaves more or less fleshy, and covered with a whitish, mealy powder. This is, perhaps, the com-

monest species ; it grows 1—3 feet high.—Fl. July,
September.   Annual.

CHENOPODIUM BONUS-HENRICUS (*Good King Henry*).

* There are several other British species of this un-
interesting family, some of which have nothing but their
rarity to recommend them, and others are remarkable
only for the tendency of their leaves to assume a triangular outline, the margin being variously lobed and

toothed. The characters of most are difficult of discrimination, so that botanists are agreed neither as to the number of species nor names.

2. ATRIPLEX *(Orache)*.

1. *A. pátula* (Spreading fruited Orache).— *Stem* spreading, often with the central branch erect; *leaves* triangular, with 2 spreading lobes at the lower angles, toothed, the *upper leaves* narrow, entire; *flowers* in tufted spikes; *perianth* of the fruit warty at the back.—Cultivated and waste ground, and on the sea-shore; abundant. A common weed, with straggling, furrowed stems, often tinged with red; distinguished from the Goose-foot family by the solitary seed being shut in between 2 triangular, leaf-like valves. The main stem is usually erect, the rest are prostrate, appearing as if they had been bent down by force.—Fl. July, August. Annual.
2. *A. laciniata* (Frosted Sea Orache).—*Stem* spreading; *leaves* with three angles, wavy at the edge, and toothed, mealy beneath. —Sea-shore, frequent. Distinguished from the preceding by its mealy leaves, and the whitish hue of the whole plant.—Fl. July, August. Annual.
   * Several other species are described by botanists, but the remark annexed to the preceding family applies equally well to this.

3. BETA *(Beet)*.

1. *B. marítima* (Sea Beet).—The only British species. —Sea-shore, common. A tall succulent plant, about 2 feet high, with large, fleshy, glossy *leaves*, angular *stems*, and numerous leafy *spikes* of green flowers, which are arranged 1 or 2 together, with a small leaf at the base of each. The root leaves when boiled are quite as good as *Spinach*.—Fl. August. Perennial.

BETA MARITIMA (*Sea-Beet*).

4. SÁLSOLA (*Saltwort*).

1. *S. Kali* (Prickly Saltwort).—The only British species.—Sandy sea-shore, common. A small plant with prostrate, branched *stems*, and succulent, awl-shaped *leaves*, each of which terminates in a sharp prickle; the *flowers* are solitary, and have 3 *bracts* at the base of each. The whole plant abounds in alkaline salt, whence its name.—Fl. July. Annual.

SALICORNIA HERBACEA (*Jointed Glasswort*).

5. SALICORNIA (*Glasswort*).

1. *S. herbacea* (Jointed Glasswort).—*Stem* herbaceous, jointed ; *leaves* 0.—Salt marshes, abundant. A singular plant 4—8 inches high, consisting of a number of fleshy joints, each of which is fitted into the one below, entirely destitute of leaves, and bearing between every two joints of the terminal branches 3 inconspicuous green flowers. —Fl. August, September. Annual.

* *S. radícans* (Rooting Glasswort) is a less common species, having prostrate, rooting *stems*. Both species abound in soda, which is used in the manufacture of glass ; hence the name *Glasswort.*

ORD. LXIX.—POLYGONEÆ.—THE PERSICARIA TRIBE.

Flowers often bearing *stamens* only, or *pistils* only. *Perianth* deeply 3—6 parted, often in two rows ; *stamens* 5— 8 from the base of the perianth ; *ovary* 1, not attached to the perianth ; *styles* 2 or 3 ; *fruit* a flattened or triangular nut.—Herbaceous plants, distinguished by the above characters and by bearing alternate leaves furnished at the base with membranous stipules, which encircle the stalk. The perianth is often coloured, and as the flowers, though not large, are numerous, and grow in spikes or panicles, many of them are handsome plants. Others, as the Dock, are unsightly weeds ; they are found in all parts of the world, from the Tropics to the Poles. The properties residing in the leaves and roots are very different, the former being acid and astringent, and sometimes of an agreeable flavour ; the latter nauseous and purgative. The powdered root of several species of *Rhéum* affords the valuable medicine Rhubarb, and the leafstalks of the same plants are much used for making tarts ; the sharp taste is attributed to the presence of oxalic, nitric, and malic acids. Two native kinds of Sorrel, and several of Dock, belong to the genus *Rumex.*

Sorrel (*R. acetósa*) is sometimes used in the same way as Rhubarb-stalks; but the species mostly employed in cookery is *R. scutáta.* To the genus *Polýgonum* belong *P. Fagopýrum* (Buck-wheat, or *Beech-wheat*), so called from the resemblance in shape between its seeds and the mast of the Beech-tree. In some countries the flour derived from its seeds is made into bread, but in England it is not much cultivated except as food for pheasants, which are very partial to it. *P. tinctórum* is extensively cultivated in France and Flanders for the sake of the blue dye afforded by its herbage, and several other species are used as medicine. *Triplaris Americana* attains the dimensions of a tree, and is remarkable for being infested by ants, which excavate dwellings for themselves in its trunk and branches. (See "Forest Trees of Britain," vol. ii. p. 65.)

1. POLÝGONUM (Persicaria).—*Perianth* deeply 5-cleft, not falling off; *stamens* 5—8; *styles* 2 or 3; *fruit* a triangular or flattened nut. (Name in Greek signifying *having many knees,* or joints, from the numerous joints of the stem.)

2. RUMEX (Dock).—*Perianth* deeply 6-cleft, in two rows, the interior segments large; *stamens* 6; *styles* 3; *fruit* a triangular nut, covered by the enlarged inner perianth. (Name, the Latin name of the plant.)

3. OXYRIA (Mountain-Sorrel).—*Perianth* deeply 4-cleft, in two rows, the interior segments large; *stamens* 6; *styles* 2; *fruit* a flattened nut with a membranous wing. (Name from the Greek *oxys,* sharp, from the acid flavour of the stem and leaves.)

1. POLÝGONUM (*Persicaria*).

* *Styles* 3; *fruit triangular.*

1. *P. Bistorta* (Bistort, Snakeweed).—*Stem* simple, erect, bearing a single dense spike; *leaves* egg-shaped, the lower ones on winged stalks.—Moist meadows, not common. A rather handsome plant, with a large twisted

root, and several stems 1—1½ foot high, each of which
bears a cylindrical spike of flesh-coloured flowers.   The
English names *Bistort* (twice twisted) and Snake-weed
were given in allusion to the form of the root.—Fl. June.
Perennial.

2. *P. viviparum* (Viviparous Bistort).—*Stem* simple,
erect, bearing a single loose spike which has in the lower
part small bulbs in place of flowers ; *leaves* very narrow,
their margins rolled back.—Mountain pastures, especially
in the Highlands of Scotland. A slender plant, 6—8
inches high, remarkable for its tendency to propagate
itself by small, red bulbs, which supply the place of
flowers in the lower part of the spike ; the flowers are
light flesh-coloured.—Fl. June, July. Perennial.

3. *P. aviculáré* (Common Knot-grass).—*Stem* branched,
prostrate; *leaves* narrow, eliptical; *flowers* axillary.—
Waste ground and road sides, abundant. A common
weed, with leaves which are furnished with chaffy stipules,
and with minute flesh-coloured or greenish-white flowers.
It varies greatly in size, and in rich soil often has a tendency to grow nearly erect.—Fl. all the summer. Annual.

* Closely allied to *P. aviculáré* are *P. Roberti*, a sea-
side plant, with stouter stems, glaucous foliage, and large
shining fruit ; and *P. maritimum*, also a sea-side plant,
but very rare, distinguished from the preceding by its
large, nerved stipules, and shrubby stem.

4. *P. Fagopýrum* (Buck-wheat).—*Stem* erect ; *leaves*
heart-arrow-shaped, acute; *flowers* in spreading panicles.
—A doubtful native, found only near cultivated land,
where it is grown principally as food for pheasants.
The stems are branched and about a foot high, the flowers
light flesh-coloured.—Fl. July, August. Annual.

5. *P. Convólvulus* (Climbing Persicaria). — *Stem*
twining ; *leaves* heart-arrow-shaped ; *segments* of the
perianth bluntly keeled ; *fruit* roughish. — Cultivated
ground, abundant. A mischievous weed with the habit
of the *Field Convólvulus*, twining round the stems of corn
and other plants, and bearing them down by its weight.

The flowers are greenish-white, and grow in axillary spikes about 4 together.—Fl. July, August.   Annual.

POLYGONUM CONVOLVULUS (*Climbing Persicaria.*)

  * *P. dumetorum* (Copse Buck-wheat), is distinguished from the last by its more luxuriant growth, its winged perianth, and shining fruit.   It grows in bushy places in the South of England.

Styles mostly 2 ; fruit flattened.

  6. *P. amphibium* (Amphibious Persicaria).—*Stem* erect, or supported in the water by the floating leaves ; *flowers* in oblong spikes ; *stamens* 5 ; *leaves* oblong,

heart-shaped at the base.—Ditches and banks of pools, frequent. So different are the forms assumed by this plant when growing in the water and on land, that the varieties might well be taken for two distinct species. In the water, the stems are 2—3 feet long, being supported by long-stalked, floating, smooth leaves ; on land, the stems are about a foot high, and the leaves narrow and rough. In both forms of the plant the spikes of flowers are rose-coloured and handsome.—Fl. July—September. Perennial.

7. *P. Persicaria* (Spotted Persicaria).—*Stem* erect, branched ; *leaves* narrow, tapering, often spotted ; *flowers* in spikes ; *stamens* 6 ; *styles* forked ; *stipules* fringed.— Waste and damp ground, abundant. A common weed 1—2 feet high, distinguished by its rather large leaves stained with purple, and numerous oblong spikes of greenish or pinkish-white flowers.—Fl. July, August. Annual.

* *P. lapathifolium* (Pale-flowered Persicaria), closely resembles the last ; it is distinguished by having 2 distinct, instead of forked *styles*, and by not having the *stipules* fringed ; in both species the *leaves* are sometimes white with silky down.

8. *P. Hydropiper* (Water-Pepper).— *Stem* erect ; *leaves* narrow, tapering ; *flowers* in loose, drooping spikes ; *stamens* 6.—Ditches and places where water has stood during winter, abundant. Well distinguished by its slender, drooping spikes of greenish flowers. The fresh juice is acrid, but not of an unpleasant flavour, and is said to cure pimples on the tongue.—Fl. August, September. Annual.

* Some botanists reckon three other species of *Polygonum*, besides the foregoing, but they are of rare occurrence and difficult to be distinguished, with the exception of *P. minus*, several stations for which are given ; it resembles *P. Hydropiper*, but is a smaller plant with upright *spikes*, narrower *leaves*, and nearly undivided *styles*.

2. RUMEX (*Dock Sorrel*).

* *Flowers having both stamens and pistils ; herbage not acid.*

1. *R. Hydrolápathum* (Great Water-Dock).—*Leaves* narrow, elliptical, tapering at both ends, the lower ones heart-shaped at the base; enlarged *segments* of the *perianth* bluntly triangular, tubercled. — River-banks, frequent. A picturesque plant 4—6 feet high, with exceedingly large leaves, and several stems which bear numerous green flowers in almost leafless whorls.—Fl. July, August. Perennial.

* There are about ten other species of Dock, some of which are rarely to be met with, others far too common. The most abundant kind is *R. obtusifólius* (Broad-leaved Dock), too well known to need any description: *R. crispus* (Curled Dock) has acute curled leaves, and is also common: *R. sanguíneus* (Bloody-veined Dock) has the veins of its leaves tinged of a beautiful crimson; the other species are less frequent.

* * *Stamens and pistils on different plants; herbage acid.*

2. *R. Acetósa* (Common Sorrel). — *Leaves* oblong, slightly arrow-shaped at the base.—Meadows, abundant. A slender plant about 2 feet high, with juicy stems and leaves, and whorled spikes of reddish-green flowers. Well known for the grateful acidity of its herbage.—Fl. June, July. Perennial.

3. *R. Acetosella* (Sheep's Sorrel).—*Leaves* tapering to a point, produced at the base into long arrow-shaped barbs.—Dry gravelly places, abundant. Much smaller than the last, and often tinged, especially towards the end of summer, of a deep red hue.—Fl. May—July. Perennial.

RUMEX ACETOSA (*Common Sorrel.*)

3. OXYRIA (*Mountain Sorrel*).

1. *O. reniformis* (Mountain Sorrel).—The only species.
—Damp places near the summits of high mountains,
frequent. Approaching the *Common Sorrel* in habit,
but shorter and stouter. The *leaves* are all from the
root, fleshy, and kidney-shaped; the *flowers* are green,
and grow in clustered spikes; the herbage has a grateful
acid flavour.—Fl. June—August. Perennial.

Ord. LXX.—ELÆAGNEÆ.—Oleaster Tribe.

Stamens and *pistils* on separate plants. *Barren flowers* in catkins ; *perianth* tubular ; *stamens* 3—8, sessile on the throat of the perianth; *fertile flower* solitary, tubular, not falling off; *ovary* 1-celled; *style* short; *stigma* awl-shaped; *fruit* a single nut, enclosed within the fleshy perianth.—Trees or shrubs, with leaves which have no stipules, but are covered with scurfy scales. They are found in all parts of the northern hemisphere. The fruit of several species of Elæagnus is eaten in the East, and the flowers are highly fragrant, and abound in honey, which, in some parts of Europe, is considered a remedy for malignant fevers. The only British species is the Sea Buckthorn (*Hippophaë Rhamnoides*).

1. Hippophaë (Sea Buckthorn).—*Stamens* and *pistils* on separate plants ; *barren flowers* in small catkins; *perianth* of 2 valves; *stamens* 4, with very short fila- ments; *fertile flowers* solitary, *perianth* tubular, cloven at the summit; *style* short; *stigma* awl-shaped; *fruit* a 1-seeded nut, enclosed in the fleshy perianth. Name of doubtful etymology.)

1. Hippophaë (*Sea Buckthorn*).

1. *H. Rhamnoides* (Sea Buckthorn, Sallow-Thorn).— The only species.—Sand-hills and cliffs on the eastern coast of England. A thorny shrub 4—5 feet high, with very narrow, silvery leaves, small greenish flowers, which appear with the leaves in May, and numerous orange-coloured berries, which are of an acid flavour, and very juicy. The stems, roots, and foliage are said to impart a yellow dye.—Fl. May. Shrub. (See Miss Pratt's " Com- mon Things of the Sea-side.")

Ord. LXXI.—THYMELEÆ.—Daphne Tribe.

Calyx tubular, coloured, 4- rarely 5-cleft, occasionally having scales in its mouth ; *stamens* 8, 4 or 2, inserted in the tube of the perianth; *ovary* 1-celled; *style* 1 ; *stigma* undivided ; *fruit* a 1-seeded nut or drupe.—Shrubs with undivided laurel-like leaves, remarkable for their tough bark, which is of a highly acrid nature, causing excessive pain if chewed, and raising a blister if applied to the skin. Both the bark and root of Mezéreon (*Daphné Mezéreon*) are used in medicine; they are of very violent effect, whether taken inwardly or applied externally. The berries of Spurge-Laurel are poisonous to all animals except birds. In the East the bark of several species is manufactured into ropes and paper. The inner bark of *Lagetta lintearia*, when macerated and cut into thin pieces, assumes a beautiful net-like appearance, whence it has received the name of Lace-bark. In the South of Europe two plants belonging to this tribe are used to dye wool yellow. The seeds of *Inocarpus édulis* are eaten when roasted, and have the taste of Chestnuts. *Daphné Japónica*, with its varieties, is commonly cultivated in conservatories and gardens for the sake of the delicious fragrance of its flowers. The only British genus belonging to this tribe is—

1. Daphne (Spurge-Laurel).—Characters given above. (Name, the Greek for a *Laurel*, which it resembles in the character of its foliage.)

1. Daphne (*Spurge-Laurel*).

1. *D. Laureola* (Spurge-Laurel).—*Flowers* in drooping, axillary clusters; *leaves* evergreen.—Woods, not unfrequent. A low shrub, about 2 feet high, very little branched, and remarkable for its smooth, erect stems, which are bare of leaves except at the summit. The

leaves are smooth, shining, and evergreen; the flowers
are green, and in mild weather fragrant; the berries,
which are egg-shaped and nearly black, are, as has been
noted above, poisonous.   From the tendency of this
plant to bear its proportionally large leaves only on the
summit of the stem, it has some resemblance to a group
of Palms.—Fl. March.   Shrub.

DAPHNE LAUREOLA (*Spurge-Laurel*).

* *D. Mezereum* is occasionally found in situations
where it is apparently wild; but it is not considered a
native; its purple, fragrant *flowers* appear before the
leaves, and are sessile on the branches; the *leaves* are
not evergreen; the *berries* red.

Ord. LXXII.—SANTALACEÆ.—Sandal-wood Tribe.

Perianth attached to the ovary, 4 or 5 cleft, valvate when in bud; *stamens* as many as the lobes of the perianth, and opposite to them; *ovary* 1-celled; *style* 1; *stigma* often lobed; *fruit* a hard, dry drupe.—The plants of this order are found in Europe and North America, in the form of obscure weeds; in New Holland, the East Indies, and the South Sea Islands as large shrubs, or small trees. Some are astringent, others yield fragrant wood. Sandal wood is the produce of *Sántalum album*, an East Indian tree, and is used both medicinally and as a perfume. In New Holland and Peru the seeds of some species are eaten. The only British plant belonging to this tribe is—

1. Thesium (Bastard Toad-flax).—Characters given above. (Name of doubtful origin.)

1. Thesium (*Bastard Toad-flax*).

1. *T. linophyllum* (Bastard Toad-flax).—The only British species.—Chalky hills, not common. A rather small plant, with a woody *root*, nearly prostrate *stems*, very narrow, pointed *leaves*, and leafy clusters of whitish *flowers*.—Fl. July. Perennial.

Ord. LXXIII.—ARISTOLOCHIEÆ.—Birthwort Tribe.

Perianth attached to the ovary below, tubular above, with a wide mouth; *stamens* 6—12 inserted on the ovary; *ovary* 3—6 celled; *style* 1; *stigmas* rayed, as many as the cells of the ovary; *fruit* 3—6 celled, many seeded.—Herbs or shrubs, often climbing, with simple leaves and solitary, axillary flowers, very abundant in the warmer parts of South America, but rare elsewhere.

The plants of this order are generally bitter, tonic and stimulant. The dried and powdered leaves of Asarabacca (*Asarum Europæum*) are used in the preparation of cephalic snuffs, exciting sneezing, and giving relief to head-ache and weak eyes. Virginian Snake-root (*Aristolóchia serpentaria*) and other allied species are used as antidotes to the bite of venomous snakes. The juice extracted from the root of a South American species is said to have the power of stupifying serpents if placed in their mouths. Other African species are said to be used by the Egyptian jugglers to stupify the snakes with which they play tricks during the exhibition of their art. The wood of Aristolochia is remarkable for not being arranged in concentric layers, but in wedges. A thin slice is a beautiful object for examination under a microscope of low power.

1. ARISTOLOCHIA (*Birthwort*).—*Perianth* tubular, curved, swollen at the base, the mouth dilated on one side; *anthers* 6, inserted on the style; *stigma* 6-lobed; *capsule* 6-celled. (Name in Greek denoting the supposed medicinal virtues of the plant.)

2. ÁSARUM (Asarabacca).—*Perianth* bell-shaped, 3-cleft; *stamens* 12 inserted at the base of the style; *stigma* 6-lobed; *capsule* 6-celled. (Name, from the Greek *a*, not, and *seira*, a wreath, denoting that the plant was by the ancients excluded from garlands.)

1. ARISTOLOCHIA (*Birthwort*).

1. *A. Clematítis* (Birthwort).—The only species found growing in wild situations in Britain, but not considered indigenous.—Woods and among ruins in the east and south of England, rare. A singular plant, with creeping *roots*, slender, unbranched, erect *stems*, and large heart-shaped *leaves*; the *flowers*, which grow several together, are of a dull yellow colour, swollen at the base, contracted above, and expanding into an oblong *lip* with a short point.—Fl. July, August. Perennial.

2. ASARUM (*Asarabacca*).

1. *A. Europœum* (Asarabacca).—The only species found in Britain, and not considered indigenous.— Woods in the North, rare. A curious plant, consisting of a very short *stem*, bearing two large shining *leaves*, and a solitary dull-green drooping *flower.*—Fl. May. Perennial.

ORD. LXXIV.—EMPÉTREÆ.—CROW-BERRY TRIBE.

Stamens and *pistils* on separate plants; *perianth* of several scales arranged in 2 rows, the inner resembling petals; *stamens* equal in number to the inner scales, and alternate with them; *ovary* of 3, 6, or 9 cells, on a fleshy disk; *style* 1; *stigma* rayed; *fruit* fleshy, with long cells; *seeds* 1 in each cell. Small heath-like ever-green shrubs, with minute axillary flowers, chiefly in-habiting Europe and North America. The leaves and fruit are slightly acid. The berries of the Crow-berry (*Empetrum nigrum*), though of an unpleasant flavour, are eaten in the Arctic regions, and are considered as a preventive of scurvy.

1. EMPETRUM (Crow-berry).—*Perianth* of 3 outer and 3 inner scales. (Name in Greek signifying growing *on a rock.*)

1. EMPETRUM (*Crow-berry*).

1. *E. nigrum* (Black Crow-berry, Crake-berry).—The only British species; abundant on mountainous heaths in the North.—A small, prostrate shrub, with the habit of a Heath. The *stems* are much branched; the *leaves* are oblong, very narrow, and have their margins so much recurved as to meet at the back, the flowers are small and purplish, growing in the axils of the upper leaves. The berries, which are black, are much eaten by moor-fowl.—Fl. May. Perennial.

Ord. LXXV.—EUPHORBIACEÆ.—Spurge Tribe.

Stamens and *pistils* in separate flowers; *perianth* lobed, with various scales or petal-like appendages; *stamens* varying in number and arrangement; *ovary* mostly 3-celled, with as many *styles* and *stigmas; fruit* generally 3-celled and 3-seeded.—A large order, very difficult to be defined even by the experienced botanist, and, therefore, very likely to puzzle the beginner, who must not be disheartened if he is a long while in reducing to their place in the system those plants belonging to it which he first meets with. Linnean botanists differ as to the Class in which the Spurges should be placed; nor is it agreed upon to which Sub-class in the Natural System they should be referred; for, though the European species have only a single *perianth*, many of the tropical genera are undoubtedly furnished with both *sepals* and *petals.* The order contains nearly 200 genera, and it is necessary to examine many of these, before the relation can be traced between those families which most differ. The number of species is thought to be not less than 2,500, which are distributed over most of the tropical and temperate regions of the globe, especially the warmer parts of America. They are either trees, shrubs, or herbs, and some kinds have the external habit of the cactus tribe. Among so numerous an assemblage of plants, we should expect to find a great dissimilarity of properties, which, indeed, exists to a certain extent; yet nearly all agree in being furnished with a juice, often milky, which is highly acrid, narcotic, or corrosive, the intensity of the poisonous property being usually proportionate to the abundance of the juice. Of the genus *Euphorbia*, Spurge, which gives name to the order, ten or twelve species are natives of Britain. The British Spurges are all herbaceous, and remarkable for the singular structure of their green flowers, and their acrid milky juice, which exudes plentifully when either the stems or leaves are wounded. A small quantity of this placed upon the tongue produces

a burning heat in the mouth and throat, which continues
for many hours. The unpleasant sensation may be
allayed by frequent draughts of milk. The roots of
several of the common kinds enter into the composition
of some of the quack fever medicines; but they are too
violent in their action to be used with safety. The Irish
Spurge is extensively used by the peasants of Kerry for
poisoning, or rather stupifying, fish. So powerful are
its effects, that a small creel, or basket, filled with the
bruised plant, suffices to poison the fish for several miles
down a river. *Euphorbia Láthyris* is sometimes, though
erroneously, called in England the Caper-plant. Its
unripe seeds are pickled, and form a dangerous substitute
for genuine capers, which are the unexpanded flower-
buds of *Cáparis spinosa*, a shrub indigenous to the most
southern countries of Europe. Among the foreign Spurges,
some species furnish both the African and American
savages with a deadly poison for their arrows. Another,
called in India *Tirucalli*, furnishes an acrid juice, which
is used in its fresh state for raising blisters. Other kinds
are used in various parts of the world as medicines, but
require to be administered with caution. The gum resin,
Euphorbium, of chemists, is procured from three species
growing in Africa and the Canaries, by wounding the
stems, and collecting in leathern bags the sap which
exudes. It is an acrid poison, highly inflammable, and so
violent in its effects, as to produce severe inflammation of
the nostrils if those who are employed in powdering it do
not guard themselves from its dust. Pliny relates that
the plant was discovered by King Juba, and named by him
after his physician, Euphorbus. The Manchineel tree
(*Hippómané Mancinella*) is said to be so poisonous, that
persons have died from merely sleeping beneath its shade.
Its juice is pure white, and a single drop of it falling
upon the skin burns like fire, forming an ulcer often
difficult to heal. The fruit, which is beautiful and looks
like an apple, contains a similar fluid, but in a milder
form; the burning it causes in the lips of those who bite

it guards the careless from the danger of eating it. *Jatropha Manihot*, or Manioc, is a shrub about six feet high, indigenous to the West Indies and South America, abounding in a milky juice of so poisonous a nature, that it has been known to occasion death in a few minutes. The poisonous principle, however, may be dissipated by heat, after which process the root may be converted into the most nourishing food. It is grated into a pulp, and subjected to a heavy pressure until all the juice is expressed. The residue, called *cassava*, requires no further preparation, being simply baked in the form of thin cakes on a hot iron hearth. This bread is so palatable to those who are accustomed to it, as to be preferred to that made from wheaten flour; and Creole families, who have changed their residence to Europe, frequently supply themselves with it at some trouble and expense. The fresh juice is highly poisonous; but, if boiled with meat and seasoned, it makes an excellent soup, which is wholesome and nutritious. The heat of the sun, even, is sufficient to dissipate the noxious properties, for if it be sliced and exposed for some hours to the direct rays of the sun cattle may eat it with perfect safety. The roots are sometimes eaten by the Indians, simply roasted, without being previously submitted to the process of grating and expressing the juice. They also use the juice for poisoning their arrows, and were acquainted with the art of converting it into an intoxicating liquid before they were visited by Europeans. By washing the pulp in water and suffering the latter to stand, a sediment of starch is produced, which, under the name of *tapioca*, is extensively imported into Europe, where it is used for all the purposes to which arrow-root and sago are applied. It is light, digestible, and nourishing, so much so, indeed, that half a pound a day is said to be sufficient to support a healthy man. Caoutchouc, or India-rubber, is a well-known elastic gum, furnished in greater or less abundance by many plants of this order, but especially by a South American tree, *Siphonia* or *Hévea elástica*.

The fragrant aromatic bark called cascarilla is produced by a shrub belonging to this order, *Croton Eleutheria*, a native of the Bahamas, and by other species of *Croton* indigenous to the West Indies and South America. Croton oil is the product of *Croton Tiglium*, and is so violent a medicine, as to be rarely administered until all other remedies have failed. Castor-oil is expressed from the seeds of *Ricinus communis*, an African tree, frequently to be met with in English gardens under the name of *Palma-Christi*, where, however, it only attains the rank of an annual herbaceous plant. The Box is the only British tree belonging to this order, of the poisonous properties of which it partakes, though to a limited extent. In some parts of Persia it is very abundant; and in these districts it is found impossible to keep camels, as the animals are very fond of browsing on the leaves, which kill them. The Dog-Mercury (*Mercuriális perennis*) is an herbaceous plant, common in our woods, and an active poison; another species, *M. annua*, is less frequently met with, and, though poisonous, is not so virulent as the other species.

1. EUPHORBIA (Spurge).—*Perianth* or *involucre* bell-shaped, containing 12 or more *barren flowers* or *stamens*, and 1 *fertile flower* or *pistil; ovary* 3-lobed; *styles* 3; *stigmas* 2-cleft; *capsule* 3-celled, 3-seeded. (Name from Euphorbus, physician to Juba, an ancient king of Mauritania, who first employed the plant as medicine.)

2. MERCURIÁLIS (Mercury).—*Stamens* and *pistils* on different plants.—*Perianth* 3-cleft to the base; *barren flower, stamens* 9, or more; *fertile flower, styles* 2; *ovary* 2-lobed; *capsule* 2-celled, 2-seeded. (Name in honour of the heathen god Mercury.)

3. BUXUS (Box).—*Stamens* and *pistils* in separate flowers, but on the same plant.—*Perianth* 4-cleft to the base; *barren flower* with 1 bract; *stamens* 4; *fertile flower* with 3 bracts; *styles* 3; *capsule* with three horns, 3-celled; *cells* 2-seeded. (Name, the Latin name of the tree.)

EUPHORBIA AMYGDALOIDES (*Wood Spurge*).

1. EUPHORBIA (*Spurge*).

* *Involucre tipped with pointed or angular glands.*

1. *E. amygdaloídes* (Wood-Spurge).—*Stem* branched above in an umbellate manner into about 5 rays; *rays* 2-forked; *bracts* perfoliate; *leaves* narrow, egg-shaped, hairy beneath; *glands* of the *involucre* crescent-shaped.

—Woods, abundant. A common woodland plant, with somewhat shrubby stems, 1—2 feet high, conspicuous in spring and summer for its golden-green leaves and flowers, and in autumn for the red tinge of its stems and leaves.—Fl. March, April. Perennial.

* To this group belong *E. Peplus* (Petty Spurge), a very common garden weed, 3—4 inches high, distinguished by its pale hue, its 3-rayed and forked umbel of numerous flowers, the involucres of which are crescent-shaped, with long horns: *E. exigua* (Dwarf Spurge), common in corn-fields, distinguished from the preceding by its narrow, glaucous leaves, and slenderer habit: *E. Parálias* (Sea Spurge), a stout, somewhat shrubby plant, growing in large masses on the sandy sea-shore, and remarkable for its numerous, imbricated, glaucous leaves: and *E. Portlándica* (Portland Spurge), also a marine species, readily distinguished from the last by its less robust habit, and the red tinge of its stems and leaves : *E. Láthyris* (Caper-Spurge), a tall, herbaceous species, 2—4 feet high, common in gardens, is remarkable for the glaucous hue of its foliage, its heart-shaped, taper-pointed bracts, and very large capsules, which abound to a great degree, as well as the rest of the plant, in the milky, acrid fluid found throughout the family.

* * *Glands of the involucre not pointed.*

2. *E. helioscópia* (Sun Spurge).—*Umbel* of 5 rays, which are often repeatedly forked ; *leaves* oblong, tapering towards the base, serrated above; *capsule* smooth.—Cultivated ground, abundant. Varying in size from a few inches to 2 feet in height, but easily distinguished by the golden-green hue of its spreading umbel, which is large in proportion to the size of the plant, and has several serrated leaves at its base.—Fl. July, August. Annual.

* *E. platyphýlla* (Warted Spurge) is a rare species, which might be mistaken for a small specimen of *E. helioscopia;* it is, however, well marked by having its leaves slightly hairy beneath, and its capsules rough, with warts at the back : *E. Peplis* (Purple Spurge) is peculiar to the sandy sea-shore; it grows quite flat on the sand, sending out several branches at right angles to the root, in a circular manner ; the whole plant is of a beautiful purple hue.   Several other species of *Spurge* are described by British botanists, but they are either very rare, or not considered indigenous.

MERCURIALIS PERENNIS (*Dog's Mercury*).

2. MERCURIÁLIS (*Mercury*).

1. *M. perennis* (Dog's Mercury). — Perennial; *stem* simple ; *leaves* stalked, roughish.—Woods and shady places, abundant.   A common, woodland, herbaceous plant; sending up from its creeping roots numerous

undivided stems about a foot high. Each stem bears

BUXUS SEMPERVIRENS (*Common Box-tree*).

several pairs of rather large, roughish leaves, and among
the upper ones grow the small, green flowers, the barren

on long stalks, the fertile sessile. — Fl. April, May. Perennial.

2. *M. annua* (Annual Mercury).—Annual; *stem* branched; *leaves* sessile, smooth.—Waste places; not common. Taller than the last, and branched. The leaves are smaller and of a light-green hue. Barren and fertile flowers are sometimes found on the same plant.—Fl. August. Annual.

3. Buxus (*Box*).

1. *B. sempervirens* (Common Box-tree).—A small tree growing in great abundance, and apparently wild, on Box-hill, Surrey, where it ripens its seeds. A dwarf variety is common in gardens. For a full description of this valuable tree, see "Forest Trees of Britain."—Fl. April. Small tree.

Ord. LXXVI.—URTICEÆ.—Nettle Tribe.

Stamens and *pistils* generally in separate flowers, and often on different plants; *perianth* divided, not falling off, sometimes wanting; *stamens* equal in number to the lobes of the perianth, and opposite to them; *anthers* curved inwards in the bud, and often bursting with elasticity; *ovary* 1, simple; *fruit* a hard and dry 1-seeded capsule.—A difficult order, the limits of which are variously assigned by different botanists. In its widest extent it contains a number of valuable fruit, as the famous Bread-fruit, and Jack-fruit, (*Artocarpus incisifolia* and *A. integrifolia*,) the Fig, Mulberry, and Sycamore of the Scriptures. The Upas-tree of Java, and *Palo-de-vaca*, or Cow-tree of Demerara, are arranged in the same Order, with many others. In its more limited extent the Nettle Tribe contains 23 families, comprising, almost entirely, rough-leaved plants, which, though they occasionally acquire the stature of trees, have, nevertheless, little more

than an herbaceous texture, their wood being remarkable
for its lightness and sponginess. They are found in most
parts of the world, occurring as weeds in the temperate
and colder regions, and attaining a larger size in hot
climates. The British species of Nettle (*Urtíca*) are
well known for the burning properties of the juice contained in the stings, with which their foliage is plentifully armed. But, painful as are the consequences of
touching one of our common Nettles, they are not to be
compared with the effects of incautiously handling some
of the East Indian species. A slight sensation of pricking
is followed by a burning heat, such as would be caused
by rubbing the part with a hot iron: soon the pain extends,
and continues for many hours, or even days, being at-
tended by symptoms such as accompany lock-jaw and
influenza. A Java species produces effects which last
for a whole year, and are even said to cause death.
Specimens of the Tree-Nettle were measured by Back-
house, in Australia, the trunks of which were found to
measure 18, 20, and 21 feet in circumference. In some
species the fibre is so strong that cordage is manufactured
from it. The burning property of the juice is dissipated
by heat, the young shoots of the common Nettle being
often boiled and eaten as a pot-herb, and the tubers of
Urtíca tuberósa are eaten as potatoes.

1. Urtica (Nettle).—*Stamens* and *pistils* in separate
flowers, on the same or different plants; *barren flower,
perianth* of 4 leaves, *stamens* 4; *fertile flower, perianth*
of 2 leaves, 1-seeded. (Name from the Latin *uro*, to
burn, from its stinging properties.)

2. Parietaria (Pellitory).—*Stamens* and *pistils* in the
same flower; *perianth* 4-cleft; *stamens* 4; *filaments* at
first curved inwards, finally spreading with an elastic
spring; *fruit* 1-seeded. (Name from the Latin *paries*, a
wall, where these plants often grow.)

3. Húmulus (Hop).—*Stamens* and *pistils* on different
plants; *barren flower, perianth* of 5 leaves; *stamens* 5;
fertile flower, a catkin composed of large, concave scales,

174   URTICEÆ.

each of which has at its base two *styles* and 1 seed.
(Name from the Latin *humus*, rich soil, in which the
plant flourishes.)

URTICA DIOICA (*Great Nettle*).

1. Urtíca (*Nettle*).

1. *U. dioíca* (Great Nettle).—*Leaves* heart-shaped at
the base, tapering to a point; *flowers* in long, branched
clusters.—A common weed, too well known to need
further description.—Fl. July, August. Perennial.

2. *U. urens* (Small Nettle).—*Leaves* elliptical ; *flowers* in short, nearly simple clusters.—Waste places, abundant. Smaller than the last, but closely resembling it in habit and properties.—Fl. July—October. Annual.

    * *U. pilulífera* (Roman Nettle) is an annual species, of local occurrence ; it bears its flowers in globular heads.

PARIETARIA OFFICINALIS (*Common Pellitory-of-the-Wall*).

2. PARIETARIA (*Pellitory-of-the-wall*).

    1. *P. officinális* (Common Pellitory-of-the-wall).—The only British species.—A much-branched, bushy, herbaceous plant, with narrow, hairy *leaves*, reddish, brittle *stems*, and small, hairy *flowers* which grow in clusters in the

axils of the leaves.   The *filaments* are curiously jointed
and elastic, so that if touched before the expansion of the
flower, they suddenly spring from their incurved position
and shed their pollen.   In rural districts an infusion of
this plant is a favourite medicine.—Fl. all the summer.
Perennial.

HUMULUS LUPULUS (*Common Hop*).

3. HÚMULUS (*Hop*).

1. *H. Lúpulus* (Common Hop).—A beautiful climbing
plant, commonly cultivated for the sake of its *catkins*,
which are used to give a bitter flavour to beer, and naturalized in many places.—Fl. July.   Perennial.

Ord. LXXVII.—ULMACEÆ.—Elm Tribe.

Stamens and *pistils* in the same or different flowers; *perianth* bell-shaped, often irregular; *stamens* equalling in number, and opposite to the lobes of the perianth; *ovary* not attached to the perianth, 2-celled; *styles* and *stigmas* 2; *fruit* 1 or 2-celled, not bursting, drupe-like, or furnished with a leafy border.—Trees or shrubs with rough leaves and clustered flowers (never in catkins), inhabiting temperate climates, and often forming valuable timber-trees. The most important genus is that of Elm (*Ulmus*), a full account of which will be found in " Forest Trees of Britain," vol. ii.

SEED OF ELM

1. Ulmus (Elm).—*Perianth* bell-shaped, 4—5-cleft, persistent; *stamens* 5; *styles* 2; *capsule* thin and leaf-like, containing a single seed. (Name, the Latin name of the tree.)

1. Ulmus (*Elm*).

1. *U. campestris* (Common small-leaved Elm).— *Leaves* obliquely wedge-shaped at the base, tapering to

a point; *fruit* oblong, deeply cloven.—Hedge-rows and parks, abundant.—Fl. March.   Tree.

ULMUS CAMPESTRIS (*Common small-leaved Elm*).

* For a further description of this and the other British species of Elm, see " Forest Trees of Britain."

ORD. LXXVIII.—AMENTACEÆ.—Catkin-bearing Tribe.

Stamens and *pistils* in separate flowers, and often on different plants; *barren flowers* in heads or catkins, composed of scales; *stamens* 1—20, inserted on the scales; *fertile flowers* clustered, solitary, or in catkins; *ovary* usually simple; *stigmas* 1 or more.—An extensive order, containing a large number of trees which are highly valued for their fruit, timber, bark, and other minor productions. They are most abundant in temperate climates, comprising a large proportion of our English Forest Trees. They have been subdivided by botanists into several sub-orders, or groups, four of which contain British specimens. The first sub-order, SALICINEÆ (the Willow group), is distinguished by bearing all its flowers in catkins, the fruit being a 2-valved capsule, containing numerous seeds tufted with down. In the sub-order MYRÍCEÆ (Sweet-Gale group), the flowers are all in catkins, and the ripe fruit assumes a drupe-like appearance, from being invested by the fleshy scales of the catkin. In BETULINEÆ (Birch group) the flowers are all in catkins, and the fruit is thin and flattened, containing 1 or 2 seeds, which are not tufted with down. In CUPULIFERÆ the fertile flowers grow in tufts or spikes, the barren flowers in catkins, and the fruit is either wholly or partially invested with a tough case, termed a *cúpula*. All the British trees belonging to this order are described at length in "Forest Trees of Britain," where mention also occurs of many foreign species, which are worthy of notice for their application to useful purposes or peculiarity of productions.

Sub-order I.—SALICINEÆ.—*Willow Group.*

1. SALIX (Willow).—*Stamens* and *pistils* on different plants (*diœcious*); *scales* of the *catkin* imbricated; *stamens* 1—5; *stigmas* 2; *capsule* of 2 valves, 1-celled; *seeds* numerous, tufted with cottony down. (Name, the Latin name of the plant.)

2. PÓPULUS (Poplar).—*Stamens* and *pistils* on different plants; *scales* of the *catkin* jagged; *stamens* 8—30; *stigmas* 4 or 8; *capsule* of 2 valves, obscurely 2-celled; *seeds* numerous, tufted with cottony down. (Name from the Latin *pópulus*, and signifying *the tree*

SALIX (*Willow*).

of the people, which it was considered to be at Rome,
and in France during the Revolutions.)

POPULUS (*Poplar*).

MYRICA (*Sweet-Gale*).

Sub-order II.—MYRÍCEÆ.—*Sweet-Gale Group.*

3. MYRÍCA (Sweet-Gale).—*Stamens* and *pistils* on different plants; *scales* of the *catkin* concave; *stamens*

4—8; *stigmas* 2; *fruit* drupe-like, 1-seeded. (Name the Greek name of the *Tamarisk.*)

BETULA (*Birch*).

Sub-order III.—BETULINEÆ.—*Birch Group.*

4. BÉTULA (Birch).—*Stamens* and *pistils* in separate flowers (*monœcious*); *scales* of the *barren catkin* in threes; *stamens* 10—12; *scales* of the *fertile catkin* 3-lobed, 3-flowered; *stigmas* 2; *fruit* flattened, 1-seeded, winged. (Name, the Latin name of the tree.)

* Two species of Birch are natives of Britain; *B. alba,* the White Birch, a forest tree, which for lightness, grace, and elegance, stands unrivalled, and has been styled "the Lady of the Woods;" and *B. nana,* Dwarf Birch, a mountain shrub, with wiry branches, and numerous round, notched leaves, which are beautifully veined.

5. ALNUS (Alder).—*Stamens* and *pistils* in separate flowers ; *scales* of the *barren catkin* 3-lobed, 3-flowered ;

ALNUS (*Alder*).

stamens 4; *scales* of the *fertile catkin* 2-flowered, permanent, becoming hard and dry; *stigmas* 2; *fruit* flattened, not winged. (Name, the Latin name of the tree.)

* *Alnus glutinósa* (Common Alder) is the only British species belonging to this family. It is a widely diffused tree, growing in swampy ground throughout most of the temperate regions of the globe.

FAGUS (*Beech*).

Sub-order IV.—CUPULIFERÆ.—*Mast-bearing Group.*

* *Stamens and pistils in separate flowers* (*monœcious*).

6. FAGUS (Beech).—*Barren flowers* in a globose catkin; *stamens* 5—15; *fertile flowers* 2 together, within a 4-lobed, prickly involucre; *stigmas* 3; *nuts* 3-cornered, enclosed in the enlarged involucre. (Name in Greek *phegos*, a species of Oak, in Latin *fagus*, a Beech.)

7. CASTÁNEA (Chestnut).—*Barren flowers* in a very long, spike-like catkin; *stamens* 10—20; *fertile flowers* 3 together, within a 4-lobed, very prickly involucre; *stigmas* 6; *nuts* not distinctly 3-cornered, enclosed in the enlarged involucre. (Name, the Latin name of the tree.)

CASTANEA (*Chestnut*).

8. QUERCUS (Oak).—*Barren flowers* in a long, drooping catkin; *stamens* 5—10; *fertile flowers* with a cup-shaped

QUERCUS (*Oak*).

scaly involucre; *stigmas* 3; *fruit* an acorn. (Name, the Latin name of the tree.)

CORYLUS (*Hazel*).

9. CÓRYLUS (Hazel).—*Barren flowers* in a long, drooping, cylindrical catkin; *scales* 3-cleft; *stamens* 8; *fertile flowers* several, enclosed in a bud-like involucre; *stigmas* 2; *nut* enclosed in the enlarged jagged involucre. (Name, the Latin name of the tree.)

10. CÁRPINUS (Hornbeam).—*Barren flowers* in a long, cylindrical catkin; *scales* roundish; *fertile flowers* in a loose catkin; *scales* large and leaf-like, 3-lobed; *stigmas*

2; *nut* strongly ribbed.   (Name, the Latin name of the tree.)

CARPINUS (*Hornbeam*).

* For a full description of the trees belonging to this order, see " Forest Trees of Britain."

ORD. LXXIX.—CONIFERÆ.—FIR TRIBE.

Stamens and *pistils* in separate flowers, and often on different trees. *Stamens* collected in sets around a common stalk ; *fertile flowers* in *cones*, destitute of styles and stigmas; *fruit* a *cone*, composed of hardened scales or bracts, bearing, at the base of each, naked seeds, which are often winged.   For a full description of this impor- tant Tribe see " Forest Trees of Britain," vol. ii.

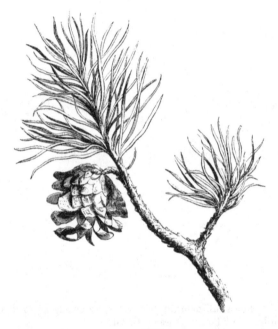

PINUS (*Fir*).

1. PINUS (Fir).—*Barren flowers* in clustered scaly catkins, the upper *scales* bearing sessile anthers ; *fertile flowers* in an egg-shaped catkin, which finally becomes a woody cone; *seeds* winged. (Name, the Latin name of the tree.)

JUNIPERUS (*Juniper.*)

2. JUNÍPERUS (Juniper).—*Barren flowers* in scaly catkins ; *anthers* attached to the base of the scales ; *fertile flowers* in catkins of a few united scales, which finally become a fleshy *berry* containing 3 seeds. (Name, the Latin name of the tree.)

* *J. communis* (Common Juniper) is a native of all the northern parts of Europe, and in Great Britain is generally found on hills and heathy downs, especially where the soil is chalky. The berries are much used to flavour hollands or geneva, a spirit distilled from corn.

TAXUS (*Yew*).

3. TAXUS (Yew).—*Barren flowers* in oval catkins which are scaly below; *stamens* numerous; *fertile flowers* solitary, scaly below ; *fruit* a naked seed, surrounded at the base by the enlarged pulpy scales. (Name, the Latin name of the tree.)

* *T. baccata* (Common Yew), the only British Yew, is an evergreen tree, remarkable for its longevity. The foliage is poisonous ; but the berries are said to be inno-cuous, being often eaten by children without ill effect. The variety called Irish Yew has erect, instead of spread-ing branches.

191

Class II. MONOCOTYLEDONOUS PLANTS.

In the plants belonging to this class the *embryo* of the seed is accompanied by a single *cotylédon*. The stem consists of *woody fibre, cellular tissue,* and *spiral vessels;* but there is no true *bark* nor *pith,* nor is the *wood* arranged in concentric layers. The stem increases in density (scarcely at all in diameter) by deposits at or near the centre; hence the plants of this class are called ENDOGENOUS, (increasing by additions on the inside.) As new substance is deposited, the old layers of wood are pressed outwards, and thus the hardest part is near the circumference. The growth of the stem is usually produced by a single terminal *bud,* without the aid of buds in the axils of the leaves; there are, however, exceptions to this rule, and the stem is often hollow. The principal *veins* of the *leaves* are parallel, not forming a complicated net-work. The *flowers* are furnished with *stamens* and *pistils,* 3 or some multiple of 3 being the predominating number of the parts of fructification. A large number are destitute of petals, the place of which is supplied by *scales* or *chaff* (glumes).

SUB-CLASS I.

PETALOIDEÆ.

Flowers furnished with petals, arranged in a circular order, or without petals.[1]

(1) Sub-class II., GLUMACEÆ, contains plants which have, instead of petals, chaffy scales or *glumes,* which are not arranged in a circular order, as is the case with the Petaloideæ, but are imbricated, such as the GRASSES and SEDGES.

ORD. LXXX.—HYDROCHARIDACEÆ.—FROG-BIT TRIBE.

Flower-buds enclosed in a sheath; *sepals* 3, green; *petals* 3; *stamens* 3, 9, 12 or more; *ovary* inferior, 1- or many-celled; *style* 1; *stigmas* 3—9; *fruit* dry or juicy, not bursting, 1- or many-celled.—A tribe of aquatic plants, often floating, among which the most remarkable is *Valisneria spirális*, the flower of which grows at the extremity of a long, spiral stalk. As the bud expands, the spire partially uncoils, allowing the flower to float on the surface for a few hours, and then contracts again, drawing the seed-vessel beneath the surface, there to ripen its seeds. The number of species is very small, two only being natives of Britain.

1. HYDRÓCHARIS (Frog-bit).—*Stamens* and *pistils* on different plants; *stamens* 9—12; *ovary* 6-celled; *stigmas* 6. (Name from the Greek *hydor*, water, and *charis*, elegance, the plants being showy aquatics.)

2. STRATIÓTES (Water-soldier).—*Stamens* and *pistils* on different plants; *stamens* about 12, surrounded by many imperfect ones; *ovary* 6-celled; *stigmas* 6. (Name, the Greek for a soldier, from its rigid, prickly, sword-shaped leaves.)

1. HYDRÓCHARIS (*Frog-bit*).

1. *H. Morsus-ranæ* (Frog-bit).—The only British species.—A floating aquatic, with creeping *stems*, round-ish, stalked *leaves*, and delicate, white flowers, which grow two or three together from a pellucid 2-leaved *sheath.*— Ponds and ditches, not general.—Fl. July, August. Perennial.

HYDROCHARIS MORSUS-RANÆ (*Frog-bit*).

2. STRATIÓTES (*Water-soldier*).

1. *S. aloídes* (Water-soldier).—The only British species; growing in ditches in the East of England.— The *roots* extend to some distance into the mud, and throw out numerous rigid prickly *leaves*, like those of an Aloe; the flower-stalk is about 6 inches high, and bears at its summit a 2-leaved sheath, containing several deli-

cate white *flowers*, bearing *stamens*, or one flower only,
bearing *pistils*. It rises to the surface before flowering,
and then sinks to the bottom.—Fl. July.   Perennial.

ORD. LXXXI.—ORCHIDEÆ.—ORCHIDEOUS TRIBE.

Sepals 3, often coloured; *petals* 3, the lowest unlike
the rest, and frequently spurred; *stamens* and *style*
united into a central column; *pollen* powdery or viscid,
sometimes raised in masses on minute stalks; *ovary* 1-
celled; *stigma* a viscid hollow in front of the column;
fruit a 3-valved capsule, with 3 rows of seeds.—A very
extensive tribe of perennial herbaceous plants, with
fibrous or tuberous roots, fleshy or leathery leaves, all the
veins of which are parallel, and flowers so variable in
form as to defy general description, yet so peculiar that
the slightest experience will enable the student to refer
them to their proper tribe.   British species have for the
most part two or more glossy sheathing leaves, and bear
their flowers in simple spikes or clusters.   The colour
of the flowers is purple, mottled with various other tints,
flesh-coloured, white, or greenish.   The structure of the
lower lip of the corolla is in many cases most singular;
sometimes resembling in form, size, and colour insects
which naturally frequent the places where the flowers
grow; such are the Bee, Fly, and Spider Orchis, (*Ophrys*
apifera, O. muscifera, and *O. aranifera.*)   In other instances the same organ presents a fantastic caricature of
some more important subject of the animal kingdom;
such are the Man, and Monkey Orchis, (*Aceras anthro-*
pophora, and *Orchis macra.*) The same mimicking extends
to foreign species:—" So various are they in form," says
Dr. Lindley, " that there is scarcely a common reptile or
insect to which some of them have not been likened."
Occasionally the structure is more complex: in *Caleàna*
nigrita the column is a boat-shaped box, resembling a
lower lip; the lip itself forms a lid that exactly fits it,

and is hinged on a claw, which reaches the middle of
the column; when the flower opens, the lip turns round
within the column and falls back, so that the flower
being inverted, it stands fairly over the latter. The
moment a small insect touches its point, the lip makes
a sudden revolution, brings the point to the bottom of
the column, and makes prisoner any insect which the box
will hold. When it catches an insect, it remains shut as
long as its prey continues to move about; but if no capture
is made, the lid soon recovers its position. The rusty
flowers of *Spiculæa ciliáta*, when spread open, may be
compared to long-legged spiders, the lip, with a long
solid plate, looking like their body, while an appendage
at its point, which is apparently moveable, is not unlike
the head of such a creature. Orchideous plants are to
be found in all climates except the very coldest and
driest; they are most abundant in the hot, damp regions
of the tropics, where they exist in the greatest profusion,
not, as in temperate countries, deriving their nourishment
from the earth, but supported by the moisture that floats
around them. Clinging to the trunks and branches of
trees, to the stems of ferns, and even to the bare rock,
they seem to adopt the habits of animals as well as to
imitate their forms. In many of these the flowers only
are conspicuous, the plant itself consisting of creeping,
claw-like roots, and tufts of elliptical bulbs, from the summit of which spring a few tough leaves, and wiry, jointed
stems, which seem incapable of producing the symmetrical
and curiously-coloured flowers they are destined shortly
to bear. Of late years, great attention has been paid to
the cultivation of exotic Orchideous plants; and by imi-
tating, as far as possible, their natural condition, great
success has been already attained. An Orchideous house
is now a common adjunct to the conservatory in the
gardens of the wealthy, where, if it be well managed,
some one or other of these curious air plants, as they
have been called, may be seen in bloom at all seasons of
the year, some clinging to broken potsherds, some to

logs of wood, some to the outer fibre of the cocoa-nut, or simply suspended by wires from the roof of the house. It is somewhat remarkable that endless as are the varieties of form which the flowers of this tribe assume, their properties vary but little. They furnish few, if any, medicines of importance; to the useful arts they contribute only a kind of cement or glue, which is recommended by no particular excellence; a nutritious substance called Salep is prepared from the roots of *Orchis máscula* and other species, but this is not extensively used; and though the flowers of many species are very fragrant, no perfume is ever extracted from them. With the exception of Vanilla, the dried fruit of *Vanilla aromatica*, which is much used in flavouring chocolate and other sweetmeats, no plant in the order can be said to be extensively used, either in the arts or sciences. Lindley conjectures the number of species to amount to 3000.

The characters by which the families of this Order are distinguished are, owing to the curious structure of the flowers comprised in it, so peculiar that they require to be attentively studied by reference to fresh specimens before any description of them can be understood. It has been thought necessary, therefore, in the case of the Orchideous Tribe, to depart from the method pursued in the other parts of this work, and instead of perplexing the student with a systematic detail of generic characters, to describe only such species as are of common occurrence, attention being paid only to their more obvious characters. The student will thus be enabled to ascertain the names of most, if not all, of the species which are likely to excite his attention. He may then examine them with accuracy, and when he has made himself acquainted with their structure and peculiarities, he will be in a position to compare whatever new species may chance to fall in his way with the descriptions given in works of higher pretension.

The first which is likely to be presented to his notice is *Orchis máscula* (Early purple Orchis), a succulent

ORCHIS MASCULA (*Early purple Orchis*).

plant about a foot high, flowering in May and June, and
abounding in woods and pastures wherever the Wild

Hyacinth flourishes. The *root* consists of two roundish solid tubers; the *leaves* are of a liliaceous texture, stained with dark purple spots, oblong and clasping the stem; the *stem* is solitary, and bears an erect cluster of purple *flowers*, mottled with lighter and darker shades; each flower rises from a somewhat twisted *ovary*, and has a long *spur*, which turns upwards. The colour of the flower, associated as it often is with Cowslips and Wild Hyacinths, is rich and beautiful; but the odour is strong and offensive, especially in the evening.

Orchis Mório (Green-winged Meadow Orchis) comes into flower about the same time with the last, and resembles it in habit. It is, however, a shorter plant, and bears fewer flowers in a cluster; it is best distinguished by the two lateral *sepals*, which are strongly marked with parallel green veins, and bent upwards so as to form a kind of hood over the column. It grows in meadows, and is often very abundant.

Orchis pyramidális (Pyramidal Orchis) bears at the summit of a somewhat slender stem a dense cluster, broad at the base and tapering to a point, of small deep rose-coloured *flowers*, which are remarkable for the length and slenderness of the spur. It usually grows on chalk or limestone, and flowers in July.

Orchis maculáta (Spotted Orchis) may be distinguished from either of the preceding by its root, which consists of two flattened tubers, divided at the extremity into several finger-like lobes. Its *leaves* are spotted like those of *O. máscula*, and its *flowers* are light purple, curiously marked with dark lines and spots. It grows abundantly on heaths and commons, flowering in June and July.

Orchis latifolia (Marsh Orchis) is a taller plant than the last, but has, like it, palmated roots; the *leaves* are remarkably erect; *flowers* rose-coloured or purple; and the *bracts*, which taper to a fine point, are longer than the flowers. It grows abundantly in marshes and wet pastures, and blossoms in June and July. All the above species, especially *O. Mório*, occasionally bear white flowers.

GYMNADENIA CONOPSEA (*Sweet-scented Orchis*).

Gymnadénia conópsea (Sweet-scented Orchis) somewhat resembles *Orchis maculáta;* the *flowers* are rose-
purple, but not spotted, and very fragrant; the *spur* is
very slender, and twice as long as the ovary. It grows
in dry, hilly, or mountainous pastures, and flowers in
June and July.

HABENARIA BIFOLIA (*Butterfly orchis*).

Habenaria bifolia (Butterfly Orchis) is a singular plant, but not appropriately named, for the resemblance which its flowers bear to a butterfly is very slight. It bears two broad *leaves* immediately above the root; the *stem* is slender and angular, about a foot high, and bears a loose cluster of greenish-white *flowers*, which are remarkable for the length of the spur, and for the strap-shaped *lower lip* of the *corolla*. It grows on heaths and the borders of woods, blooming in June. The flowers are fragrant in the evening. *H. viridis* (Green Habenaria) and *H.*

álbida (Small White Habenaria) are small plants, from 6 to 8 inches high, the former with green *flowers,* the latter with *flowers* which are white and fragrant.

LISTERA OVATA (*Twayblade*).

Listera ováta (Twayblade) grows from 12 to 18 inches high, and is well marked by its bearing, about half way up its cylindrical stem, two opposite egg-shaped *leaves;* the *flowers* are small and green.   It is not uncommon in woods and orchards, and flowers in June.   *Listera cordáta* is a much smaller plant, with two heart-shaped leaves.

It occurs in mountainous districts, and flowers from June to August.

Listera Nidus-avis (Bird's nest) is a pale reddish-brown plant, about a foot high, entirely destitute of *leaves*, the place of which is supplied by numerous sheathing, brown *scales*. The *root* consists of many short fleshy fibres, from the extremities of which the young plants are produced. It is found sparingly in shady woods, flowering in June.

NEOTTIA SPIRALIS (*Lady's Tresses*).

Neottia spirális (Lady's Tresses) is a curious little plant, from 4 to 6 inches high, with tuberous *roots* an a spike of small white *flowers*, which are arranged in a single row, and in a spiral manner, in some specimens from left to right, in others from right to left, round the upper portion of the stalk. The flowers are fragrant in the evening. The *leaves* form a tuft just above the crown of the root, but do not show themselves until the flowers have begun to expand; they are remarkably tenacious of life, continuing to unfold even while subjected to the pressure made on the blotting-paper during the process of drying. Not uncommon in dry pastures, flowering in September and October.

Ophrys apífera (Bee Orchis).—The distinctive character of the flowers of this curious plant is given in its name; and the same may be said of *O. muscífera* (Fly Orchis); both species occur in considerable abundance in many of the limestone and chalk districts. No one who has heard that plants exist bearing these names, can doubt their identity, should they fall in his way. They flower in July. The Spider Orchises, *Ophrys arachnítes* and *O. aranífera*, are of rare occurrence.

Epipactis grandiflóra (White Helleborine) grows from 1 foot to 2 feet high, bearing several rather broad *leaves* on the stem, and a loose spike of large, pure white *flowers*. It grows in woods, being almost if not entirely confined to a chalky soil, and flowers in June. *Epipactis latifolia* (Broad-leaved Helleborine) and *E. palustris* (Marsh Helleborine) are of similar habit; the former bearing a long loose spike of greenish-purple *flowers;* the latter a spike of *flowers* variegated with purplish-green, white and rose-coloured.

The above description includes all the Orchideous plants which are of common occurrence. The rarer species indigenous to Britain are:—*Orchis ustuláta* (Dwarf dark-winged Orchis), *O. fusca* (Great brown-winged Orchis), *O. militaris* (Military Orchis), *O. macra* (Monkey Orchis), and *O. hircína* (Lizard Orchis).—These grow only in chalk districts.

Goódyera repens (Creeping Goodyera) is a small plant

OPHRYS APIFERA (*Bee Orchis*).

with creeping roots, growing in fir woods in Scotland;
Corallorhíza innáta (Coral-root) is well marked by its
curiously-toothed *roots*, which in figure resemble branched
coral; it occurs in marshy woods in Scotland. *Áceras
anthropóphora* (Man Orchis) bears a long, loose spike of
greenish yellow *flowers;* it grows in dry chalky places.

Malaxis paludósa(Bog Orchis), the smallest British Orchideous plant, 2—4 inches high, grows in spongy bogs, and bears a spike of minute green *flowers*. *Líparis Loesélii* (Two-leaved Líparis) is confined to the eastern counties, where it is rarely found in spongy bogs; it bears a spike of 6—12 yellowish *flowers* on a triangular stalk. *Cypripedium Calcéolus* (Lady's Slipper), distinguished by its large inflated lip, occurs but rarely in the woods of the north of England, and is pronounced by Sir W. J. Hooker "one of the most beautiful and interesting of our native plants."

Ord. LXXXII.—IRIDACEÆ.—Iris Tribe.

Perianth 6-cleft; *stamens* 3, rising from the base of the sepals; *ovary* inferior, 3-celled; *style* 1; *stigmas* 3, often petal-like; *capsule* 3-celled, 3-valved; *seeds* numerous. Principally herbaceous plants with tuberous or fibrous roots, long, and often sword-shaped, sheathing leaves, and showy flowers, which seldom last a long time. Chiefly natives of warm and temperate regions, and most abundant at the Cape of Good Hope, where, at the time of its discovery by the Portuguese, the natives mainly supported themselves on the roots of the plants of this tribe, together with such shell-fish as were left on shore by the receding tide. *Iris, Crocus, Ixia*, and *Gladíolus* are favourite garden flowers; *Iris pseud-ácorus* (Yellow Iris, or Flag) is one of our most showy marsh plants. Few species are used in the arts or sciences; the roots of *Iris Florentína* afford Orris-root, which, when dried, has a perfume resembling that of Violets, and is used as an ingredient in tooth-powder; the dried stigmas of *Crocus sativus*, were anciently much prized as a dye, and are still employed for the same purpose, as well as in medicine and cookery; and the roots of a few species are used by barbarous nations as an occasional article of food.

1. Iris.—*Perianth* with the 3 outer divisions longer,

IRIS PSEUD-ACORUS (*Yellow Iris, Flag, Corn Flag*).

and reflexed; *stigmas* 3, petal-like, covering the stamens. (Name from Iris, the rainbow, from the beautiful colouring of the flowers.)

2. TRICHONÉMA.—*Perianth* in 6 equal, spreading divisions; *tube* shorter than the *limb; stigma* deeply 3-cleft, its lobes 2-cleft, slender. (Name from the Greek *thrix*, a hair, and *néma*, a thread or filament.)

3. CROCUS.—*Perianth* in 6 equal, nearly erect divisions;

tube very long; *stigma* 3-cleft, its lobes inversely wedge-shaped. (Name from the Greek *crocos*, saffron, and that from *crocé*, a thread.)

1. Iris (*Flower-de-luce*).

1. *I. Pseud-ácorus* (Yellow Iris, Flag, Corn Flag).— *Leaves* sword-shaped; *perianth* not fringed, its inner divisions smaller than the stigmas.—Marshes and banks of rivers, common. A stout aquatic plant, with creeping, acrid roots, sword-shaped leaves 2—3 feet long, and large, handsome, yellow flowers. The root yields a black dye, and the roasted seeds, it is said, may be used as a substitute for coffee.—Fl. June, July. Perennial.

2. *I. fœtidissima* (Stinking Iris).—*Leaves* sword-shaped; *perianth* not fringed, its inner divisions about as large as the stigmas; *stem* slightly flattened.—Woods and hedges in the west and south-west of England, not uncommon. Resembling the last in habit, but smaller. The flowers are of a dull leaden hue, and the leaves so acrid as to leave a burning taste in the mouth, or even to loosen the teeth. The whole plant, when bruised, emits a disagreeable odour. Sir W. J. Hooker states that "in Devonshire it is so frequent, that one can hardly avoid walking among it when herbalising, and being annoyed by the smell." This, however, is an exaggeration. The berry-like seeds, which are of a beautiful scarlet colour, remain attached to the plant all through the winter.—Fl. June—Aug. Perennial.

* Many species of Iris are cultivated for their beauty, some of which are occasionally found apparently wild, in the neighbourhood of gardens.

2. Trichonéma.

1. *T. Columnæ* (Columna's Trichonéma).—The only British species.—A small bulbous plant, 3—4 inches high, with very narrow *leaves*, and solitary purplish *flowers* tinged with yellow, partaking the characters of the Iris and Crocus. It grows only on a sandy pasture called the Warren, at Dawlish, Devon.—Fl. March, April. Perennial.

CROCUS SATIVUS (*Saffron Crocus*).

3. Crocus.

1. *C. sativus* (Saffron Crocus).—*Leaves* appearing after the flowers, linear; *flower-stalks* enveloped with a double sheath; *stigma* long and drooping.—Said to be naturalized at Saffron-Walden, in Essex, where it is largely cultivated

for the sake of the *saffron* afforded by its dried stigmas, the only part of the plant which is used. The flowers are purple.—Fl. September. Perennial.

* Several other species of *Crocus* are sometimes found in the neighbourhood of gardens. In some of these the leaves and flowers appear together in spring. The seed-vessels, in their early stage, are concealed, at the base of the long tube, beneath the ground; but when the flowers are withered, the stalk rises and exposes them to the action of the air and sun, to be ripened.

Ord. LXXXIII.—AMARYLLIDACEÆ.—Amaryllis Tribe.

Perianth of 3 coloured *sepals* and 3 *petals; stamens* 6, arising from the sepals and petals, sometimes united by the base of their filaments; *ovary* inferior, 3-celled; *style* 1; *stigma* 3-lobed; *fruit* a many-seeded capsule, or a 1—3-seeded berry.—An extensive tribe, principally composed of herbaceous plants with bulbous roots, sword-shaped leaves, and showy flowers, which are distinguished from the true lilies by their *inferior* ovary; that organ in the *Lily Tribe* being *superior*, and enclosed within the corolla. Large and beautiful species belonging to this Order are found in abundance in Brazil, the East and West Indies, and especially the Cape of Good Hope. In the temperate regions they are less common, and by no means so showy. In Great Britain, it is doubtful whether a single species is indigenous, though the number of varieties cultivated in gardens, both in conservatories and in the open air, is very great. The bulbous roots of many plants belonging to the Amaryllis Tribe are poisonous; some, it is said, to such a degree that dele-terious properties are communicated to weapons dipped in their juice. The roots of the Snowdrop and Daf-fodil are emetic, and the flowers of the last (*Narcissus Pseudo-Narcissus*) are a dangerous poison. The roots of some species, however, are nutritious, affording a kind of arrowroot. From the juice of a kind of *Agávé* (*A. Americana*) a fermented liquor is made, which, under the name of "pulque," is in Mexico a common beverage. This plant, called by the Mexicans "maguey," is cultivated over an extent of country embracing 50,000 square miles. In the city of Mexico alone the consumption of pulque amounts to the enormous quantity of eleven millions of gallons, and a considerable revenue from its sale is derived by government. The plant attains maturity in a period varying from eight to fourteen years, when it flowers; and it is during the stage of flowering alone that the juice is extracted. The central stem, which encloses the flower-bud, is then cut off near the bottom, and a cavity or basin is discovered, over which the leaves are drawn close, and tied. Into this reservoir the juice distils, which otherwise would have risen to nourish and support the flower. It is removed three or four times during the 24 hours, yielding a quantity of liquor varying from a quart to a gallon and a half. The juice is extracted by means of a syphon made of a species of gourd, and deposited in bowls. It is then placed in earthern jars, and a little old pulque is added, when it soon ferments, and is immediately ready for use. The fermentation occupies two or three days, and when it ceases it is in fine order. Old pulque has an unpleasant odour, which has been compared to that of putrid meat: but when fresh it is brisk and sparkling. In time even Europeans prefer it to any other liquor. This Agávé is popularly known in England by the name of American Aloe. It grows but slowly in this climate, and, as it rarely attains perfection, is believed by many persons to flower once in a hundred years. A fine specimen which flowered at Clowance, Cornwall, in September 1837, is thus described:—" A sturdy stem 17½ inches in diameter rises from a *chevaux de frise* of spiked leaves to the height of 14 feet without a branch; but copiously beset with leaves of the same character, which gradually diminish in size, and are continued to the very top, till from being four or five feet long, they become mere bracts, measuring not as many lines. The lower branches are not, as they are often represented in engravings, shorter than those above them, but extend to the length of several feet, describing a graceful curve. A short distance above, but not so close as to appear crowded, or so exactly opposite as to present a formal appearance, another branch leaves the main stalk, a very little shorter than the lower one, and is succeeded by others, in all thirty-four, the last being twenty-five feet from the ground. Each of these bears at its extremity a mass of green flowers, the summits of which form nearly a plane surface. The number of flowers on the five lower branches amounts to 911, so that the whole, including those on the summit of the stem, must be computed, on a fair average, at no less than 5000 perfect flowers. The quantity of honey which they contained was very great, and as it was constantly dropping moistened the ground beneath. The plant, exhausted by the unusual effort, for which it had been for many years preparing, died the same year." The roots and leaves of the species of *Agáve* contain woody fibre (*pita thread*), useful for various purposes; this is separated by bruising and steeping in water, and afterwards beating. The Mexicans also make their paper of the fibres of Agáve leaves laid in layers. The expressed juice of the leaves is also stated to be useful as a substitute for soap.

1. NARCISSUS (Daffodil).—*Perianth* tubular at the base, terminating in a bell-shaped *crown* or *nectary*, which has 6 equal *sepals* and *petals* at its base. (Named after *Narcissus*, a fabulous youth, said to have been changed into a flower.)

2. GALANTHUS (Snowdrop).—*Perianth* bell-shaped; *sepals* 3 (white), spreading; *petals* 3, erect, notched. (Name in Greek signifying " milk-flower.")

3. LEUCÓJUM (Snow-flake).—*Perianth* bell-shaped, of 6 equal *sepals* and *petals*, which are thickened at the point. (Name in Greek signifying " a white violet.")

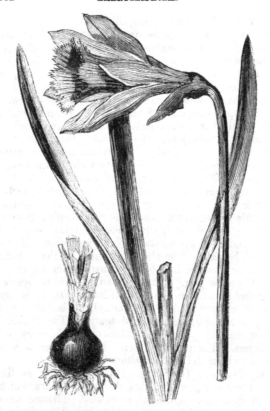

NARCISSUS PSEUDO-NARCISSUS (*Common Daffodil, Lent Lily*).

1. NARCISSUS (*Daffodil*).

1. *N. Pseudo-Narcissus* (Common Daffodil, Lent Lily).—*Flower-stalk* hollow, 2-edged, bearing near its summit a membranous sheath and a single flower; *nectary* notched and curled at the margin, as long as the sepals

and petals.—Woods and orchards, common. A favourite flower with country children, owing to its large size and showy, yellow colour; but its smell is unpleasant, and the whole plant possesses poisonous properties.—Fl. March, April. Perennial.

* Several other species of *Narcissus* are occasionally found near houses, but they are invariably the outcast of gardens.

GALANTHUS NIVALIS (*Snowdrop*).

2. GALANTHUS (*Snowdrop*).

1. *G. nivális* (Snowdrop).—The only species, and too well known to need any description. Though apparently wild in many places, it is not considered a native plant.— Fl. January—March. Perennial.

LEUCOJUM ÆSTIVUM (*Summer Snowflake*).

3. Leucójum (*Snowflake*).

1. *L. æstívum* (Summer Snowflake).—A doubtful
native, found occasionally in moist meadows in many
parts of England.—A bulbous plant about 2 feet high,
with narrow, keeled *leaves,* and 2-edged flower-stalks,
which bear each an umbel of rather large white *flowers,*

the *sepals* and *petals* of which are tipped with green.   It is a common garden plant.—Fl. May.   Perennial.

Ord. LXXXIV.—DIOSCOREACEÆ.—Yam Tribe.

Stamens and *pistils* on different plants (*diœcious*); *perianth* 6-cleft; *stamens* 6, arising from the base of the perianth; *ovary* inferior, 3-celled; *style* deeply 3-cleft; *fruit* a dry, flat capsule, or (in *Tamus*, the only British species) a berry.—Twining shrubs or herbs, approaching in habit some of the Dicotyledonous Orders, the leaves being decidedly stalked, and having netted veins; the flowers are small with 1—3 bracts each, and grow in spikes.   The Order is a small one, and is, with the exception of *Tamus* (Black Bryony) confined to tropical regions.   Dioscórea, the plant from which the Order takes its name, has large tuberous roots, which, under the name of "Yams," form as important an article of food in tropical countries as the Potato in temperate climates.   When growing it requires a support, like the Hop.   There are several varieties; the best are white and mealy, but some are yellow and watery, with a slightly bitter taste.

1. Tamus (Black Bryony). — Characters described above.   (Name, the Latin name of the plant.)

1. Tamus (*Black Bryony*).

1. *T. commúnis* (Black Bryony). — The only British species.—*Root* a large, solid *tuber*, black externally; *stem* slender, twining among bushes to the length of many feet, and clothed with numerous varnished, heart-shaped *leaves*, and clusters of small, green *flowers*, which are succeeded by elliptical scarlet *berries*.   The leaves are reticulated with veins like those of plants belonging to Dicotyledonous plants, but they are not jointed to the stem.   Late in autumn they turn dark purple or bright yellow, when, assisted by the scarlet berries, they make

a very showy appearance. In winter they die down to
the ground.—Fl. May—July. Perennial.

TAMUS COMMUNIS (*Black Bryony*).

ORD. LXXXV.—TRILLIACEÆ.—HERB-PARIS TRIBE.

Sepals and *petals* 6—8, coloured or green; *stamens*
6—10; *anthers* very long, their cells, one on each side
of the filament; *ovary* superior, with 3—5 cells, and as
many *styles; fruit* a 3—5-celled berry; *seeds* numerous.—

A small Order, containing about thirty herbaceous plants with tuberous roots, whorled, netted leaves, and large solitary, terminal flowers. They grow in the woods of temperate climates, and, like the plants of the last Order, bear some resemblance to Dicotyledonous plants. The structure of the seed, however, and the fact that the leaves are not jointed to the stem, fix them in the class of Endogenous or Monocotyledonous plants. Their properties are acrid and narcotic.

1. PÁRIS (Herb-Paris).—*Sepals* and *petals* 8, very narrow; *stamens* 8—10. (Name from the Latin *par, paris,* equal, on account of the unvarying number of the leaves.)

PARIS QUADRIFOLIA (*Four-leaved Herb-Paris, True-Love-Knot*).

1. PÁRIS (*Herb-Paris*).

1. *P. quadrifolia* (Four-leaved Herb-Paris, True-Love-Knot).—The only British species ; growing in damp woods, not common. A singular plant, with a *stem* about a foot high, bearing near its *summit* four large, pointed *leaves*, from the centre of which rises a solitary large, green *flower.*—Fl. May. Perennial.

Ord. LXXXVI.—LILIACEÆ.—Lily Tribe.

Calyx 0 ; *corolla* of 6 petals, distinct, or united into a tube ; *stamens* 6, inserted into the petals, opening inwards ; *ovary* superior, not united with the petals, 3-celled, many-seeded ; *style* 1 ; *stigma* simple or 3-lobed ; *capsule* 3-celled, 3-valved, oblong ; *seeds* numerous, flattened horizontally.—An extensive family of plants, of which the majority are herbaceous, with bulbous roots, and showy flowers ; some, however, attain the dimensions of shrubs, or even trees, in which case they resemble the Palms rather than exogenous trees, the trunk being destitute of true bark and pith, and the leaves being never jointed to the stem. Butcher's Broom (*Ruscus*) is the only British species which assumes a shrubby character ; *Asparagus* is a branching herbaceous plant, with creeping roots, scaly stems, and bristle-like leaves ; *Convallaria* (Lily of the Valley) has also creeping roots. These three produce a berry-like fruit. All other British species have bulbous roots, and the fruit when ripe is a dry capsule. Plants of the Lily tribe are most abundant in temperate climates, but attain their greatest magnitude in the tropics. *Dracæna Draco* (Dragon's Blood) is said to grow to the height of 70 feet, with a stem upwards of 40 feet in circumference. In the Canaries one is in existence which is known to have been an ancient tree in the year 1406. The leaves of many species contain a tough fibre, which is used as a substitute for hemp or flax. Among these the most remarkable is *Phormium*

tenax (New Zealand Flax).   The genus *Allium* (Onion,
Garlic, and Leek,) supplied food to the early inhabitants of
Egypt, and had divine honours paid to it.   In Kamt-
schatka, Tartary, and the Sandwich Islands, various
species are cultivated for the same purpose.   The bud
and tender part of the stem of the Grass-tree, a native
of Tasmania, is said to be nutritious, and of an agreeable
flavour, and in our own country the young shoots of
Asparagus rank among the most delicate of our esculent
vegetables.   In medicine many species are of great value,
among which, aloes, the condensed juice of *Alóé vulgáris*
&c., and squills, an extract of *Scilla marítima,* are well
known.   As ornamental plants the beauty of the Lily
Tribe has been for ages proverbial; *Lilium Chalcedónicum,*
which covers the plains of Syria with its scarlet flowers,
is said to have been the plant to which our Blessed
Saviour alluded in his Sermon on the Mount, under the
title of " the lilies of the field."   The innumerable
varieties of Hyacinth are derived from an Eastern plant,
Hyacinthus Orientális; and the Tulip (*Túlipa*) was long
the most highly prized among florists' flowers, and furnished
in Holland a subject for the most absurd speculation.

Group I. ASPARAGEÆ.—*The Asparagus Group.*

Roots never bulbous ; fruit, a berry.

1. ASPARAGUS.—*Corolla* deeply 6-cleft, bell-shaped ;
stamens 6, distinct ; *stigmas* 3, bent back.   (Name, the
Greek name of the plant.)

2. RUSCUS (Butcher's Broom).—*Corolla* deeply 6-cleft ;
stamens and *pistils* on different plants (*diœcious*); *stamens*
connected at the base; *style* surrounded by a nectary.
(Name, " anciently *bruscus,* from *bruskelen;* in Celtic,
box-holly."—*Sir W. J. Hooker.*)

3. CONVALLARIA (Lily of the Valley, Solomon's Seal).
—*Corolla* 6-cleft, bell-shaped ; *stamens* 6, distinct ;
stigma 1.   (Name from the Latin *convallis,* a valley,
the usual locality of this family.)

Group II. SCILLEÆ.—*The Squill Group.*

Root bulbous; fruit a capsule; stalk leafless.

4. HYACINTHUS (Hyacinth).—*Corolla* deeply 6-cleft,
tubular, bell-shaped ; *lobes* of the *corolla* reflexed at the
extremity. (Name from Hyacinthus, a youth fabled to
have been changed into a flower.)

5. SCILLA (Squill).—*Corolla* of 6 spreading petals;
flowers blue or purple, in a cluster or corymb, not enclosed in a sheath, falling off as the seed ripens. (Name,
the Latin name of the plant.)

6. ORNITHÓGALUM (Star of Bethlehem).—Like *Scilla,*
except that the petals are white, and do not fall off.
(Name from the Greek *ornis,* a bird, and *gala,* milk.
This plant is supposed by Linnæus to be the "dove's
dung" mentioned in 2 Kings vi. 25.)

7. ALLIUM (Garlic).—*Corolla* of 6 spreading petals ;
flowers in an umbel, at the base of which is a sheath of
1 or 2 leaves. (Name, the Latin name of the plant.)

Group III. TULIPEÆ.—*The Tulip Group.*

Like the Squill group, except that the flower-stalk is leafy.

8. GÁGEA.—*Flowers* in an umbel or corymb; *petals*
6, without a nectary; *anthers* erect, attached to the
filaments by their bases; *style* conspicuous. (Named in
honour of Sir Thomas Gage.)

9. TÚLIPA (Tulip).—*Flowers* solitary, rarely 2 on a
stem ; *petals* and *anthers* as in Gágea ; *style* 0. (Name
from *toliban,* the Persian name for a turban.)

10. FRITILLARIA (Fritillary). — *Flowers* solitary ;
petals 6, with a nectary at the base of each ; *anthers*
attached above their bases ; *style* 3-cleft at the summit.
(Name from the Latin *fritillus,* a dice-box, the common
accompaniment of a *chequer-board,* which the marking
of the flower resembles.)

* Besides the above, several other genera are described

by British botanists, of which *Anthéricum* (Spider-wort) is the only real native; it is very rare, growing only near the summit of Snowdon and other mountains in Wales.

ASPARAGUS OFFICINALIS (*Common Asparagus*).

1. ASPARAGUS.

1. *A. officinális* (Common Asparagus).—The only British species, occurring sparingly on several parts of the sea-coast, especially near the Lizard Point, Cornwall; it differs only in size from the cultivated plant. See " A Week at the Lizard."—Fl. July, August. Perennial.

RUSCUS ACULEATUS (*Butcher's Broom, Knee Holly*).

2. RUSCUS (*Butcher's Broom*).

1. *R. aculeátus* (Butcher's Broom, Knee Holly).—The only British species, and only British shrub of Endogenous growth.—Waste and bushy places, not uncommon, especially in the south of England. A low shrub 3—4 feet high, with erect green *stems*, which are branched and plentifully furnished with very rigid *leaves*, terminating each in a sharp spine. The *flowers* are minute, greenish-white, and grow singly from the centres of the

leaves; the *berries* are as large as marbles, round, and of a brilliant scarlet colour.—Fl. April, May. Shrub.

3. CONVALLARIA (*Lily of the Valley*).

1. *C. majális* (Lily of the Valley).—*Leaves* all from the root; *flowers* drooping, in a long, one-sided cluster.— Woods, in a light soil, not frequent. A common, and universally admired garden plant, equally prized for its globular, pure white flowers, and for its delicious perfume. —Fl. May. Perennial.

CONALLARIA MULTIFLORA (*Solomon's Seal*).

2. *C. multiflóra* (Solomon's Seal).—*Stem* roundish, bearing numerous elliptical *leaves*, which are all turned

one way, and opposite them small clusters of drooping *flowers,* which are all turned the other way; *filaments* hairy.—Woods, in several parts of England and Scotland,

HYACINTHUS NON-SCRIPTUS (*Wild Hyacinth, Blue-bell*).

but not frequent. A singular plant, 1—2 feet high, bearing numerous whitish flowers, with green tips, shaped like old-fashioned round seals.—Fl. June. Perennial.

* *C. Polygonátum* (Angular Solomon's Seal) differs from the last species in having an angular *stem*, mostly solitary *flowers*, and smooth *filaments : C. verticilláta* bears its leaves 3—4 in a whorl. Both these species are rare.

4. HYACINTHUS (*Hyacinth*).

1. *H. non-scriptus* (Wild Hyacinth, Blue-bell).—The only British species. Too abundant and well known to need any description. The name *Hyacinthus* was originally given to some species of Lily into which the youth Hyacinthus was fabled to have been changed by Apollo. The petals are marked with dark spots, arranged so as to resemble the Greek word AI, alas! The present species, however, having no such characters on its petals, was named *non-scriptus*, not written. It is sometimes, though incorrectly, called Hair-bell; the true Hair-bell being *Campánula rotundifolia.*—Fl. May, June. Perennial.

5. SCILLA (*Squill*).

1. *S. verna* (Vernal Squill).—*Flowers* in a corymb; *bracts* narrow; *leaves* linear, appearing with the flowers. —Sea-coast in the west and north of England. A lovely little plant, 3—4 inches high, with *corymbs*, or flat clusters of blue, star-like *flowers*. The turfy slopes of the sea-coast of Cornwall are in many places as thickly studded with these pretty flowers as inland meadows are with Daisies. In a few weeks after flowering no part of the plant is visible but the dry capsules, containing black, shining seeds.—Fl. May. Perennial.

2. *S. autumnális* (Autumnal Squill).—*Flowers* in an erect cluster; *bracts* 0; *leaves* appearing after the

flowers.—Dry pastures, especially near the sea, not common. About the same size as the last, but less beautiful. Flowers, purplish blue.—Fl. August—October. Perennial.

SCILLA VERNA (*Vernal Squill*).

ORNITHOGALUM PYRENAICUM (*Spiked Star of Bethlehem*).

6. ORNITHÓGALUM (*Star of Bethlehem*).

1. *O. Pyrenáicum* (Spiked Star of Bethlehem).—
Flowers in a long spiked cluster.—Woods, not common ;
very abundant in the neighbourhood of Bath, where the
spikes of unexpanded flowers are often exposed for sale
as a pot-herb. A bulbous plant, with long, narrow

leaves, which wither very early in the season, and a leafless stalk about 2 feet high, bearing a long erect cluster of greenish-white flowers.—Fl. June, July.

* *O. umbellátum* (Common Star of Bethlehem), though not an English plant, is not unfrequently found in the neighbourhood of houses. It bears large, pure white *flowers*, which are green externally, and grow in *umbels*, or, rather, *corymbs*, opening only in sunny weather. It is a common garden plant.

ALLIUM URSINUM (*Broad-leaved Garlick, Ramsons*).

7. ALLIUM (*Garlic*).

1. *A. ursínum* (Broad-leaved Garlic, Ramsons).— *Leaves* broad and flat; *flower-stalk* triangular; *flowers*

in a flat umbel.—Woods and thickets, common. The leaves of this plant are scarcely to be distinguished from those of the Lily of the Valley; the flowers are white and pretty, but the stench of the whole plant is intolerable.—Fl. May, June. Perennial.

* Seven other species of Garlic are described by British botanists, but none of them are so common as the last. *A. Schœnóprasum* (Chives) is a pretty plant, with dense heads of bright purple flowers. In a wild state its foliage is scanty, but under cultivation becomes very abundant, in which state it is a favourite cottage pot-herb. Several other species are remarkable for bearing small bulbs among the flowers.

8. GÁGEA.

1. *G. lutea* (Yellow Gágea).—The only British species. —Woods and pastures, rare. A bulbous plant 6—8 inches high, with long narrow *leaves*, and umbels of yellow *flowers*.—Fl. March—May. Perennial.

9. TÚLIPA (*Tulip*).

1. *T. sylvestris* (Wild Tulip).—The only British species.—Chalk-pits, rare. A bulbous plant with very narrow *leaves*, and a solitary, yellow, drooping *flower*, which is fragrant, and much smaller than the garden Tulip.—Fl. April. Perennial.

10. FRITILLARIA (*Fritillary*).

1. *F. Meleágris* (Fritillary, Snake's-head).—The only British species.—Meadows and pastures in the east and south of England, not common. A bulbous plant, with very narrow *leaves*, and a solitary drooping *flower*, shaped like a Tulip, and curiously chequered with pink and dull purple.—Fl. April. Perennial.

FRITILLARIA MELEAGRIS (*Fritillary, Snake's-head*).

Ord. LXXXVII.—MELANTHACEÆ.

Meadow-saffron Tribe.

Calyx and *corolla* alike, coloured, in 6 pieces, or united below into a tube; *stamens* 6; *anthers* turned outwards; *ovary* 3-celled; *style* deeply 3-cleft; *capsule* divisible into 3 pieces; *seeds*, each contained in a membranous case.—A small tribe, containing some plants approaching the Lilies in habit, and others, the Crocuses, confined to no particular countries, but most frequent in the northern hemisphere. Many species possess acrid and poisonous properties, and are used to destroy vermin. *Cólchicum* (Meadow Saffron) is used as a specific for the gout, but it is considered a dangerous medicine.

COLCHICUM AUTUMNALE (*Meadow Saffron*).

1. Cólchicum (Meadow Saffron).—*Corolla* with a very long tube, rising from a sheath. (Name from Colchis, a country famous for medicinal herbs.)

2. TOFIELDIA (Scottish Asphodel).—*Corolla* of 6 petals; *flowers* each from a small 3-lobed sheath. (Name in honour of Mr. Tofield, an English Botanist.)

1. CÓLCHICUM (*Meadow Saffron*).

1. *C. autumnále* (Meadow Saffron).—The only British species. — Meadows, not common. A not unfrequent garden plant, with large broad *leaves*, which wither away in summer, after which appear several light purple flowers, resembling Crocuses in all respects except that they have 6 instead of 3 stamens. At the time of flowering the seed vessels are concealed beneath the ground, where they remain until the following spring, when they rise above the surface and are ripened. When grown in gardens the bulbs are often taken up as soon as the leaves have withered, and placed in a window, where they will flower without earth or water.—Fl. September, October. Perennial,

2. TOFIELDIA (*Scottish Asphodel*).

1. *T. palustris* (Mountain Scottish Asphodel).—The only British species.—Boggy ground in the north. A small plant 4—6 inches high, with tufts of narrow, sword-shaped *leaves*, and egg-shaped, almost stalkless spikes of small, yellowish *flowers*. — Fl. July, August. Perennial.

ORD. LXXXVIII.—JUNCACEÆ.—RUSH TRIBE.

Calyx and *corolla* alike, of 6, usually chaffy, pieces (sometimes coloured, as in Asphodel); *stamens* 6, inserted into the base of the petals and sepals, or sometimes 3, inserted into the sepals; *anthers* turned inwards; *ovary* superior; *style* 1; *stigmas* 3 (in Asphodel 1); *capsule* 3-valved, usually many seeded.—A tribe of marsh or bog plants, with cylindrical, or flat leaves sometimes filled with pith ; the flowers are usually small, and of a brownish green hue, but in Asphodel they are bright yellow, and the leaves are sword-shaped, like those of the Iris.

The true Rushes (Juncus) are, for the most part, social plants, and are often of considerable use in fixing the soil of marshes and bogs. The stems of the common species are used for making mats and the wicks of candles. The tall aquatic plant usually called the Bulrush belongs to the *Sedge Tribe*, the Club-rush to the Order *Typhaceæ*, and the Flowering Rush to the Order *Butomaceæ*.

1. JUNCUS (Rush) —*Perianth* chaffy; *filaments* smooth; *stigmas* 3 ; *capsule* 3-celled, 3-valved ; *seeds* numerous. (Name, the Latin name of the plant, and that from *jungo*, to join, the stems having been woven into cordage.)

2. LÚZULA (Wood-rush) —Like *Juncus*, except that the capsule is 1-celled, and only 3-seeded. (Name supposed to have been altered from the Italian *lucciola*, a glowworm, from the sparkling appearance of the heads of flowers when wet with rain or dew.)

3. NARTHECIUM (Asphodel).—*Perianth* of 6 coloured sepals and petals; *stamens* downy; *stigma* 1; *capsule* 3-celled ; *seeds* numerous. (Name from the Greek *narthex*, a rod, to which, however, the only British species bears little resemblance.)

1. JUNCUS (*Rush*).

* *Stems cylindrical, tapering to a point; leaves none.*

1. *J. effúsus* (Soft Rush).—*Stems* not furrowed ; *panicle* below the summit of the stem, branched and spreading ; *capsule* blunt.—Marshy ground, common. This and the following species are well known as the

rushes of which mats and the wicks of candles are made.
—Fl. July.   Perennial.

JUNCUS EFFUSUS (*Soft Rush*).

2. *J. conglomeratus* (Common Rush).—*Stems* not furrowed ; *panicle* below the summit of the stem, crowded;
capsule ending in a point.—Marshy places, common.
Only distinguished from the last by its dense panicle of
flowers, and pointed capsule.—Fl. July.   Perennial.

3. *J. glaucus* (Hard Rush).—*Stems* deeply furrowed, rigid; *panicle* below the summit of the stem, branched and spreading.—Marshy places and road-sides, common. Very distinct from the two last, from which it may be distinguished by its more slender, furrowed, glaucous stems, and its very loose panicle of slender flowers.—Fl. July. Perennial.

* Several other species belong to this group, but none are common except *J. maritimus* (Lesser Sea-Rush), which differs from those already described in having the portion of the stem which rises above the panicle dilated at the base so as to resemble a *bract*; it grows in salt marshes: *J. acutus* (Great Sea-Rush), the largest British species, grows on the sandy sea-shore in great abundance in a few places; it is well marked by its stout and rigid habit, and by its large, polished *capsules*.

** *Stems leafless; leaves all from the root.*

4. *J. squarrósus* (Heath Rush).—*Leaves* rigid, grooved; *panicle* terminal.—Moors and heaths, abundant. Well marked by its rigid stems and leaves, of which the latter have mostly one direction. The stems are about 1 foot high; the flowers larger than in the marsh species, and variegated with glossy brown, and yellowish white.—Fl. June, July. Perennial.

*** *Stems leafy; leaves cylindrical, or but slightly flattened, jointed internally.*

* The most common species in this group are:—
J. acutiflorus (Sharp-flowered jointed Rush), a slender plant, 1—2 feet high, with slightly flattened *stems* and *leaves*, and terminal panicles of brown sharp-pointed *flowers: J. lampocarpus* (Shining-fruited jointed Rush), resembling the last, but distinguished by its large, brown, glossy *capsule: J. obtusiflórus* (Blunt-flowered jointed Rush), rather smaller than *J. acutiflorus*, and well distin-

guished by its blunt *flowers:* and *J. uliginosus* (Lesser
Bog jointed Rush) a small and very variable plant, 3—8
inches high, bearing a few clusters rather than panicles
of flowers. All these are common in boggy ground.

JUNCUS ULIGINOSUS (*Lesser Bog jointed Rush*).

**** *Stems leafy; leaves not cylindrical nor jointed.*

* In this group there are but two common species;
J. compressus (Round-fruited Rush), a slender plant,
about a foot high ; the *leaves* are linear, and grooved
above; the *stem* is slightly flattened, and terminates in
a panicle of greenish-brown flowers ; the *capsule* is
nearly round, with a point : and *J. bufonius* (Toad

Rush), a very small species, 4—6 inches high, with repeatedly forked *stems*, and solitary green *flowers*, which
grow mostly on one side of the stem. For several other
species growing among the mountains in the north, the
student is referred to " Hooker's British Flora."

2. Lúzula (*Wood-Rush*).

1. *L. sylvática* (Great Wood-Rush).—*Leaves* hairy;
panicle spreading, much branched; *flowers* in clusters of
about 3.—Woods, abundant. A common woodland
plant, with more of the habit of a *Grass* than a Rush.
The leaves are flat and clothed with long, scattered,
white hairs; the stalk rises to the height of about 2 feet,
and bears a terminal loose cluster of brownish flowers,
with large light yellow anthers.—Fl. May, June. Perennial.

2. *L. pilósa* (Hairy Wood-Rush).—*Leaves* hairy;
panicle little branched; *flowers* solitary.—Woods, not
unfrequent. Smaller than the last, and well distinguished
by its solitary flowers, the stalks of which are bent back
when in fruit.—Fl. May, June. Perennial.

3. *L. campestris* (Field Wood-Rush).—*Leaves* hairy;
panicle of 3 or 4 dense, many-flowered clusters.—Pastures,
common. Much smaller than either of the preceding.
This is one of the first grass-like plants to show flower
in spring, when it may be distinguished from all other
meadow-herbs by its dense *clusters* or *spikes* of brownish-
green flowers, each of which contains 6 large, light-yellow
anthers.—Fl. March—May. Perennial.

* Other British species of Wood-Rush are *L. Fórsteri*
(Forster's Wood-Rush), the *panicle* of which is slightly
branched, and bears its *flowers* solitary; each *capsule* contains 3 *seeds*, having a straight tail at their summits;
it resembles *L. pilosa* in habit, but is much smaller; the
seeds of the latter plant are furnished with a long hooked
tail: *L. spicáta* (Spiked Mountain Wood-Rush) is about
the same size as *L. campestris;* it has narrow *leaves,*

LUZULA CAMPESTRIS (*Field Wood-Rush*).

bears its flowers in a compound, drooping spike, and
grows only on high mountains: *L. arcuáta* (Curved
Mountain Wood-Rush) is a small and very rare species,
found only on the summit of the Scottish mountains;
it bears its flowers in panicles, 3—5 together, on drooping
stalks.

3. NARTHECIUM (*Asphodel*).

1. *N. ossífragum* (Bog Asphodel).—The only British
species.—An elegant little plant, 6—8 inches high, with

tufts of narrow, sword-shaped *leaves*, like those of the *Iris*, and a tapering spike of star-like yellow *flowers*.

NARTHECIUM OSSIFRAGUM (*Bog Asphodel*).

The name *ossifragum*, bone-breaking, was given to this plant from its being supposed to soften the bones of cattle that fed on it. Other plants have had the same properties assigned to them, but there is little doubt that in every case the diseases in question are to be traced to the noxious exhalations from the bogs in which the plants grow, rather than to the plants themselves.— Fl. July—September. Perennial.

BUTOMUS UMBELLATUS (*Flowering Rush*).

Ord. LXXXIX.—BUTOMACEÆ.—Flowering Rush Tribe.

Sepals 3, green; *petals* 3, coloured; *stamens* varying in number; *ovaries* superior, 3, 6, or more, distinct, or

united into a mass; *carpels* many-seeded.—A small
tribe of aquatic plants with sword-shaped leaves and
conspicuous flowers. The only British example is the
Flowering Rush, described below.

1. Bútomus (Flowering Rush).—*Stamens* 9; *carpels*
6. (Name, from the Greek *bous*, an ox, and *temno*, to
cut; because cattle feeding on the leaves are liable to
cut their mouths.)

1. Bútomus (*Flowering Rush*).

1. *B. umbellatus* (Flowering Rush).—The only British
species.—A tall, aquatic plant, growing in stagnant
water and slow rivers, not uncommon. The *leaves* are
sword-shaped, 2—4 feet long, and spring all from the
root; the *flowers* are large, rose-coloured, and handsome,
and grow in a simple umbel at the top of a round stalk,
which rises several feet above the surface of the water.—
Fl. June, July. Perennial.

Order XC.—ALISMACEÆ.—Water-Plantain Tribe.

Sepals 3, green; *petals* 3, coloured; *stamens* varying
in number; *ovaries* superior, numerous; *carpels* numerous, 1- or 2-seeded.—A small tribe of aquatic plants,
often floating, with long-stalked leaves, and flowers
which in some respects resemble those of the Crowfoot
Tribe. Like the Crowfoots, too, their juice is acrid,
though the roots of some species, deprived of their acridity
by drying, are said to be used as food.

1. Alisma (Water-Plantain).—*Flowers* containing
both stamens and pistils; *stamens* 6; *carpels* numerous,
1-seeded. (Name, the Greek name of the plant, and
that said to be derived from the Celtic *alis*, water.)

2. Actínocarpus (Star-fruit).—Like *Alisma*, except
that the *carpels* are 2-seeded, and spread in a radiate
manner. (Name in Greek having the same meaning as
the English name.)

3. Sagittaria (Arrow-head).—*Stamens* and *pistils* in separate flowers (*monœcious*); *stamens* numerous; *carpels* numerous, 1-seeded. (Name from the Latin *sagitta*, an arrow, from the shape of the leaves.)

ALISMA PLANTAGO (*Great Water-Plantain*).

1. Alisma (*Water-Plantain*).

1. *A. Plantágo* (Great Water-Plantain).—*Leaves* all from the root, broad below, and tapering to a point ;

flowers in a compound, whorled panicle.—Margins of rivers, lakes, and ponds; common.   A stout, herbaceous plant, 2—3 feet high, with large, stalked leaves, ribbed like those of a *Plantain*, and a leafless, whorled panicle of lilac flowers, the petals of which are very delicate, and soon fall off.—Fl. June—August.   Perennial.

2. *A. ranunculoídes* (Lesser Water-Plantain).—*Leaves* narrow, and tapering at both ends; *flowers* in umbels.— Peaty bogs, not uncommon.   Much smaller than the last, and well marked by the above characters, as well as by its larger flowers.

* *A. natans* (Floating Water-Plantain) is found only in mountain lakes; it has floating leafy *stems,* and the *flowers,* which are solitary, are white, with a yellow spot.

2. Actínocarpus (*Star-fruit.*)

1. *A. Damasonium* (Common Star-fruit).—The only British species.—Ditches in the midland counties, not common.   An aquatic plant, with the habit of a Water-Plantain.   The *leaves* grow on long stalks and float on the surface of the water; the *flowers,* which grow in whorls, are white, with a yellow spot at the base of each petal; the *fruit* is composed of 6 pointed carpels, which are arranged in the form of a star.—Fl. June, July. Perennial.

3. Sagittária (*Arrow-head*).

1. *S. sagittifolia* (Common Arrow-head).—The only British species.—Rivers and ditches, not unfrequent. A pretty plant, well distinguished by its large arrow-shaped *leaves,* and whorled panicles of delicate, flesh-coloured *flowers,* both of which rise 6—8 inches out of the water.   This is one of the very few plants which neither smoke nor buildings have driven out of London, there being still large beds of it in the Thames, near the Temple Gardens and Hungerford Market, where the

eager botanist may even yet gather fine specimens.—Fl.
July—September.   Perennial.

SAGITTARIA SAGITTIFOLIA (*Common Arrow-head*).

Ord. XCI.—JUNCAGINACEÆ.—Arrow-grass Tribe.

Flowers perfect; *sepals* and *petals* alike, green and
small;   *stamens* 6;   *ovaries* 3—6, superior, united or
distinct; *carpels* 3—6, 1—2 seeded.—A small order of
marsh plants, with linear leaves, all proceeding from the
root, and spike-like clusters of inconspicuous flowers ;
found in many parts of the world, and possessing no
remarkable properties.

1. Triglochin (Arrow-grass).—*Flowers* in a spike ;
sepals and *petals* 6 ; *stamens* 6.   (Name from the Greek
treis, three, and *glochis*, a point : from the three points of
the capsule.)

TRIGLOCHIN PALUSTRE (*Marsh Arrow-grass*).

1. TRIGLÓCHIN (*Arrow-grass*).

1. *T. palustre* (Marsh Arrow-grass).—*Fruit* linear, of 3 combined *carpels*.—Marshy places, frequent. A plant with something of the habit of *Plantago marítima*, from which it may easily be distinguished by its fewer flowers and slenderer spike, as well as by the different structure of the flowers. The leaves are linear and fleshy.—Fl. June—August. Perennial.

2. *T. maritimum* (Sea Arrow-grass).—*Fruit* egg-shaped, of 6 combined *carpels*.—Salt marshes, common. Like the last, but well marked by its rounded, not linear capsule.—Fl. May—September. Perennial.

* *Scheuzeria palustris*, which belongs to this order, is a very rare plant, found only in the north. It has a few semicylindrical, blunt leaves, and a leafless stalk, terminating in a cluster of 5—6 green flowers.

ORD. XCII.—TYPHACEÆ.—REED-MACE TRIBE.

Stamens and *pistils* separate, but on the same plant (*monœcious*); *flowers* in dense spikes or heads, not enclosed in a sheath; *perianth* composed of 3 scales or a tuft of hairs; *stamens* 3—6, distinct, or united by their filaments; *anthers* long, and wedge-shaped; *ovary* single, superior, 1-celled; *style* short; *stigma* linear, lateral; *fruit* 1-celled, 1-seeded, not opening, angular by mutual pressure.—Herbaceous plants, growing in marshes or ditches, with jointless stems, sword-shaped leaves, and small flowers, which are only conspicuous from their compact mode of growth. The order contains only two families, examples of both which are of common occurrence in Great Britain.

1. TYPHA (Reed-Mace).—*Flowers* in spikes. (Name from the Greek *typhos*, a marsh, where these plants grow.)

2. SPARGANIUM (Bur-reed). — *Flowers* in globular heads. (Name in Greek denoting a little band, from the ribbon-like leaves.)

1. TYPHA (*Reed-Mace*).

1. *T. latifolia* (Great Reed-Mace, or Cat's Tail).— *Leaves* nearly flat; *barren* and *fertile* spikes continuous. —Ponds, common. Our largest herbaceous aquatic, often growing 6—8 feet high, with linear leaves, and stout, cylindrical stems, surmounted by a club-like spike, the lower part of which contains fertile flowers only, the upper barren. It is often, but incorrectly, called *Bulrush*, the true Bulrush being *Scirpus palustris*, a plant

which has more of the habit of a gigantic rush.—Fl.
July, August. Perennial.

TYPHA LATIFOLIA (*Great Reed-Mace, or Cat's Tail*).

2. *T. angustifolia* (Lesser Reed-Mace, or Cat's Tail).
—*Leaves* grooved below; *barren* and *fertile spikes* slightly
interrupted.—Ponds, less frequent than the last, from
which it differs by the above characters, and by its
smaller size.—Fl. July, August.   Perennial.

SPARGANIUM RAMOSUM (*Branched Bur-reed.*)

2. SPARGANIUM (*Bur-reed*).

1. *S. ramósum* (Branched Bur-reed).—*Leaves* triangular at the base, with concave sides; *stem* branched.
—Ditches, common. A large aquatic, which at a distance might be mistaken for a Flag (*Iris Pseud-ácorus*). The leaves are sword-shaped, and the flowers are collected into globular heads, of which the lower contain fertile flowers only, the upper barren.—Fl. July, August. Perennial.

2. *S. simplex* (Unbranched upright Bur-reed).—*Leaves* triangular at the base, with flat sides; *stem* unbranched.—Ditches, common. Smaller than the last, and at once distinguished by the above characters.

* *S. natans* (Floating Bur-reed) is found only in the north. It has very long, pellucid, floating *leaves*, and *flowers* resembling those of the preceding species, except that the *barren head* is usually solitary.

ORD. XCIII.—ARACEÆ.—THE CUCKOO-PINT FAMILY.

Stamens and *pistils* separate, but on the same plant (*monœcious*); *flowers* arranged on a *spadix*, or central column, and enclosed in a sheath; *perianth* 0; *stamens* numerous, sessile on the spadix; *ovaries* the same, below the stamens; *stigma* sessile; *fruit* a berry.—A curious tribe of plants, all more or less resembling the British species, *Arum maculátum*, abounding in tropical countries, and possessing acrid, or even poisonous qualities, which, however, may be dissipated by heat. The most remarkable plant of the order is the Dumb-Cane of the West Indies, a species growing as high as a man, and having the property, when chewed, of swelling the tongue and destroying the power of speech. The effects continue for several days, and are accompanied with much pain. Other species which are scarcely less unfit for food in their fresh state, are extensively cultivated in tropical countries, and produce tuberous roots, which, when cooked, are important articles of food. Even the British example of the order (*Arum maculátum*), though its juice is so intensely acrid that a single drop will cause a burning taste in the mouth and throat, which continues for hours, has roots which, when properly prepared, are wholesome and nutritious. This plant is cultivated in the Isle of Portland, and the starch procured from the roots is, under the name of Portland Sago, used as a substitute for arrow-root. Several species have been observed to evolve a considerable quantity of heat from the spadix at the time of the expansion of the sheath.

1. ARUM (Cuckoo-pint).—*Flowers* on a club-shaped *spadix*, which is naked above, and enclosed in a convolute *sheath*. (Name, the Greek name of the plant.)

1. ARUM (*Cuckoo-pint*).

1. *A. maculátum* (Cuckoo-pint, Wake-Robin, Lords-and-Ladies).—The only British species.—Hedges and woods, common in most parts of England. A succulent, herbaceous plant, with large, glossy, arrow-shaped leaves, which are often spotted with dark purple. The upper part of the *spadix* is club-shaped, and of a light pink, dull purple, or rich crimson colour, which is easily rubbed off; about the middle of the *spadix* is a ring of glands, terminating in short threads, the use of which is unknown, below this is a ring of sessile *anthers*, and yet lower down another row of sessile *ovaries*. The upper part of the *spadix* soon falls off, leaving the *ovaries*, which finally become a cylindrical mass of scarlet *berries*, which are conspicuous objects when all the rest of the plant has withered and disappeared. The *spadix* with its *sheath* may be discerned wrapped up in the young leaf-stalks, even before the leaves have risen above the surface of the ground.—Fl. May, June. Perennial.

ARUM MACULATUM (*Cuckoo-pint, Wake-Robin, Lords-and-Ladies*).

ORD. XCIV.—ORONTIACEÆ.—SWEET-SEDGE TRIBE.

Flowers perfect, arranged on a central column or *spadix*, at first enclosed in a *sheath; perianth* of 4—8 scales; *stamens* equalling the scales in number; *ovary* superior; *fruit* a berry.—A tribe of plants nearly allied to the *Araceæ*, and resembling them in properties. *Calla Æthiópica* is, under the name of Egyptian Lily, perhaps better known than the only British species, *Ácorus Cálamus*, or Sweet Sedge. This last plant is said to have supplied the "rushes" with which, before the use of carpets had been introduced into England, it was customary to strew the floors of the great. As it did not grow in the neighbourhood of London, but had to be fetched at considerable expense from Norfolk and Suffolk, one of the charges of extravagance brought against Cardinal Wolsey was that he caused his floors to be strewed with rushes too frequently. It is still used to strew the floor of the cathedral at Norwich on festival-days.

1. Ácorus (Sweet Sedge).—*Sheath* leaf-like, not convolute, overlapping the *spadix*. (Name in Greek denoting

ACORUS CALAMUS (*Sweet Sedge*).

that the plant has the power of curing diseases in the pupil of the eye.)

1. Ácorus (*Sweet Sedge*).

1. *A. Cálamus* (Sweet Sedge).— The only British species.—Watery places in Norfolk and Suffolk. An aquatic plant, with somewhat of the habit of a *sedge* or large *grass*. It is easily distinguished from all other British plants by its peculiar *spadix*, and the fragrance of its roots, stems, and leaves.—Fl. June. Perennial.

Ord. XCV.—PISTIACEÆ.—Duck-weed Tribe.

Minute floating plants, composed of simple or lobed *leaves*, and fibrous *roots*, which are not attached to the soil, propagating themselves principally by offsets, but sometimes producing on the edge of the leaves 1—2 *stamens*, and 1—4 seeded *ovaries*, enclosed in small *sheaths*. *Lemna* (Duck-weed) is the only British example, and the number of foreign species is but small.

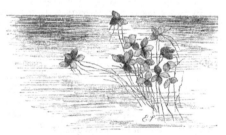

LEMNA MINOR (*Lesser Duck-weed*).

1. Lemna (*Duck-weed*).

1. *L. minor* (Lesser Duck-weed).—A minute plant, but often so abundant as to cover the surface of stagnant water, where, with the insects which it harbours, it is

greedily devoured by ducks. In this species the *leaves*
are egg-shaped, and bear each a single *root.* Three other
species have been found in Britain, for a description of
which the student is referred to " Hooker's British
Botany."

ORD. XCVI.—NAIADACEÆ.—POND-WEED TRIBE.

Submersed or floating aquatics with very cellular *stems*
and peculiar *leaves,* which are sometimes almost leathery,
but more frequently thin and pellucid. The flowers are
small, olive-green, resembling in structure the Arrow-
grasses; sometimes solitary, but more frequently arranged
in spikes. They inhabit ponds and slow streams, or
rarely salt marshes. One British species, *Zostéra marína,*
grows in the sea.

1. POTAMOGÉTON (Pond-weed).—*Flowers* in a spike ;
stamens and *pistils* in the same flower; *perianth* of 4
sepals ; *stamens* 4 ; *carpels* 4, sessile. (Name from the
Greek *pótamos,* a river, and *geíton,* a neighbour.)

2. RUPPIA.—*Flowers* about 2 on a stalk ; *stamens*
and *pistils* in the same flower ; *perianth* 0; *stamens* 4;
carpels 4, at first sessile, afterwards raised each on a
long stalk. (Named in honour of H. B. Ruppius, a
botanist of the 18th century.)

3. ZANNICHELLIA (Horned Pond-weed). — *Flowers*
axillary; *stamens* and *pistils* separate (*monœcious*);
stamen 1; *carpels* 4. (Named in honour of J. J. Zanni-
chelli, a Venetian botanist.)

4. ZOSTÉRA (Grass-wrack).—*Flowers* composed of
stamens and *pistils* alternately arranged in 2 rows in a
long leaf-like *sheath.* (Name from the Greek *zoster,* a
girdle, which the leaves resemble in form.)

1. POTAMOGÉTON (*Pond-weed*).

1. *P. natans* (Floating Pond-weed).—*Upper leaves*
elliptical, ribbed, and cellular, *lower* submersed, linear.

—Ponds and ditches, common. An aquatic plant, with cord-like stems, proportioned to the depth of the water in which it grows; smooth, floating leaves, on long stalks; and cylindrical spikes of small green flowers, which project above the surface of the water. The upper, or floating leaves are 2—3 inches in length, the lower, which are not always present, are very narrow and a foot long, or more.—Fl. June—August. Perennial.

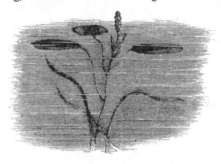

POTAMOGETON NATANS (*Floating Pond-weed*).

2. *P. perfoliatus* (Perfoliate Pond-weed).—*Leaves* all submersed, egg-shaped, embracing the stem, pellucid, 7-nerved.—Ponds and lakes, common. Remarkable for its brown, almost transparent leaves, 2—3 inches long, which when dry have the appearance of gold-beater's skin, and are so sensitive of moisture, that they will curl when laid on the palm of the hand.—Fl. June—August. Perennial.

3. *P. pusillus* (Small Pond-weed).—*Leaves* linear, very narrow; *flowers* in a long-stalked, loose spike.—Ponds and lakes, common. A tangled mass of thread-like stems, and dull, olive-green leaves, with numerous spikes of brownish flowers, which are either submersed, or partially rise above the surface of the water.—Fl. June—August. Perennial.

* From 18 to 20 species of Pond-weed are described

as natives of Britain; they all, more or less, resemble the above in habit, and as they are by no means an interesting family of plants, easy to obtain, or pleasant to examine, it is not thought necessary to describe their characters in an elementary work like this.

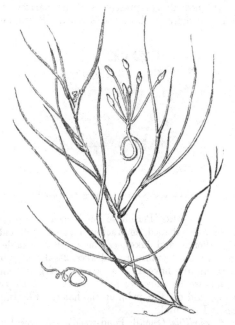

RUPPIA MARITIMA (*Sea Ruppia*).

2. RUPPIA.

1. *R. maritima* (Sea-Ruppia).—The only species, growing in salt-water ditches, distinguished from *Potamogéton pusillus* by its spiral *flower-stalks*, and long-stalked *fruit*.—Fl. July—August.

3. ZANNICHELLIA (*Horned Pond-weed*).

1. *Z. palustris* (Horned Pond-weed).—The only British species. A submersed aquatic, with the habit of *Potamogéton pusillus,* from which it may be well distinguished by its small, almost sessile, axillary *flowers,* the *stigmas* of which are unevenly cup-shaped.—Fl. August—September. Perennial.

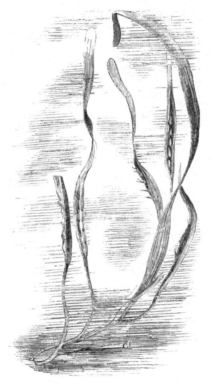

ZOSTERA MARINA (*Grass-wrack*).

4. Zostéra (*Grass-wrack*).

1. *Z. marína* (Grass-wrack).—A submersed marine aquatic, with long, cord-like *stems*, and bright green, grass like *leaves*, some of which serve as *sheaths* to the bead-like rows of small simple flowers.   The dried leaves and stems are used as beds, and are also employed in packing glass.—Fl. July, August.   Perennial.

ENGLISH INDEX

AND

GLOSSARY OF BOTANICAL TERMS.

———◆———

LATIN INDEX.

THE END.

Printed in the United States
By Bookmasters